国家林业和草原局普通高等教育"十四五"规划教材

高等院校园林与风景园林专业系列教材

康复花园设计中英双语教程

A Bilingual Textbook on Healing Garden Design

郑　丽　　陈利平　主编

中国林业出版社
China Forestry Publishing House

内 容 简 介

随着城市化进程，环境污染日益严重、都市人群压力过大等一系列问题频发，在建设美丽中国、健康中国的大背景下，作为人类文明与自然结合产物的园林环境，其生态、保健的功能越来越受重视，尤其是森林园林康养师职业被纳入新版职业大典，从园林环境解决健康问题的康复花园已经被越来越多的卫生诊疗机构和社会大众所认可。本教材围绕康复花园的内涵和外延展开阐述，包括康复花园的概念、发展沿革、理论基础、设计基础，以及不同功能型康复花园的案例应用等，均用中文和英文进行论述和讲解，并重点介绍了康复花园的中国园林文化溯源。

本教材编写人员涉及风景园林、园艺学、植物学、建筑学、心理学、医学等学科领域，可以作为风景园林专业、园林专业、园艺专业、建筑学专业、心理学等专业本科生教材，以及领域相关教师、工程师或其他从业人员的参考用书。

图书在版编目（CIP）数据

康复花园设计中英双语教程／郑丽，陈利平主编
. -- 北京：中国林业出版社，2025.1
国家林业和草原局普通高等教育"十四五"规划教材
高等院校园林与风景园林专业系列教材
ISBN 978-7-5219-2563-0

Ⅰ.①康… Ⅱ.①郑… ②陈… Ⅲ.①花园-园林设计-高等学校-教材-汉、英 Ⅳ.①TU986.2

中国国家版本馆 CIP 数据核字（2024）第 018866 号

策划编辑：康红梅
责任编辑：康红梅
责任校对：苏 梅
封面设计：北京钧鼎文化传媒有限公司
封面摄影：康红梅
————————————

出版发行：中国林业出版社
　　　　　（100009，北京市西城区刘海胡同 7 号，电话 010-83223120、83143551）
电子邮箱：jiaocaipublic@163.com
网　　址：https://www.cfph.net
印　　刷：北京盛通印刷股份有限公司
版　　次：2025 年 1 月第 1 版
印　　次：2025 年 1 月第 1 次印刷
开　　本：850mm×1168mm　1/16
印　　张：16.25
字　　数：383 千字
定　　价：59.00 元

《康复花园设计中英双语教程》
编写人员

主　编　郑　丽　　陈利平

副主编　吴祥艳　　袁惠燕　　张吟松

编写人员（按姓氏拼音排序）

白　晶（西安科技大学高新学院）

包晓鹏（云南省楚雄彝族自治州农业科学院）

陈　飞（海南大学）

陈利平（云南农业大学）

陈立人（苏州农业职业技术学院）

陈友华（云南农业大学）

陈志荣（云南农业大学）

陈志欣（山东华龙园林工程有限公司）

付晓渝（苏州大学）

黄印星［Hortian Consultancy Pte. Ltd.（新加坡）］

李房英（福建农林大学）

李茜玲（长三角先进材料研究院）

李　燕（国家植物园）

石杉杉（北京合众和管理咨询有限公司）

孙旻恺（苏州科技大学）

孙志勇（江苏省扬州五台山医院）

谭凯欣［Hortian Consultancy Pte. Ltd.（新加坡）］

乌云巴根(华侨大学)

王　胜(重庆市风景园林科学研究院)

王晓蕾(云南农业大学)

吴广平(云南农业大学)

吴祥艳(中央美术学院)

袁惠燕(苏州大学)

袁晓梅(华南理工大学)

张吟松(云南农业大学)

赵　萍(云南农业大学)

郑　丽(苏州大学)

主　审　(按姓氏拼音排序)

李树华(清华大学)

姚　雷(上海交通大学)

序 1

　　人类自六七百万年前诞生以来，99%的时间都是在绿色环境中度过的，所以在人们身体中形成了"绿色基因"，看到绿色心情就宁静，植物与自然还能促进人类的健康。这就是园艺疗愈、园林康养等绿色康养的根本所在。

　　快速化城市进程中出现的生境破碎化、城市污染严重、生物多样性降低等问题，导致"城市病"与"城市人群病"大流行。"城市病"表现在交通拥挤、住房紧张、供水不足、能源紧缺、环境污染、秩序混乱，以及物质流、能量流的输入、输出失去平衡，需求矛盾加剧等；"城市人群病"表现在亚健康人群增大、慢性患者群年轻化、青少年心理疾病增加以及手机成瘾症明显等。据研究报道，70%的城市人群患者可以通过绿色环境的人为干预得到防治，特别是在防未病、康复以及慢性病、老年性疾病、精神性疾病的疗愈与康复方面，园艺疗愈与园林康养将发挥巨大作用。

　　为了顺应当前形势与未来发展的需求，人力资源和社会保障部于2022年7月公布了新修订的《中华人民共和国职业分类大典》（简称新版《大典》），其中第四大类"社会生活与生活服务人员"下设中类"4-14（GBM41400）健康、体育和休闲服务人员"，该种类包括从事健康咨询、医疗临床、康复矫正、公共卫生、体育健康、康养休闲等辅助服务工作的人员共7个小类，并在小类"4-14-06（GBM41406）康养休闲服务人员"中，新增职业"4-14-06-01 森林园林康养师"，后者包括森林康养师、园林康养师2个工种在内的森林园林康养师成为国家认可的正式职业。

　　同时，面向园林康养职业化时代的到来，我们必须做好以下各项工作，为园林康养师的诞生与发展奠定基础：①深入开展园艺疗法与园林康养科学研究；②制定园林康养师职业标准；③加强相关院校园林康养人才培养；④建立严格的园林康养师认证体系；⑤建立高质量培训实习基地，并完善认证体系；⑥为园林康养师提供就职岗位；⑦进行园林康养师业务督导；⑧进一步完善包括园艺康养、园林康养、森林康养以及农业康养在内的绿色康养师体系。

　　《康复花园设计中英双语教程》正是上述工作的重要组成部分。该教材围绕康复花园的内涵和外延展开阐述，包括康复花园的概念、发展沿革、理论基础、设计基础，以及不同功能型康复花园的案例应用等，均用中文和英文进行论述和讲解，并重点突出了康复花园的中国园林文化溯源，这种中西兼容的教材编写不仅能培养学生的国际视野，同时也在很大程度上体现了中国特色和文化自信。在思想性、科学性和先进性方面，均能够满足教学需要。该教材将对我国园林康养与园艺疗愈事业的发展起到推动作用。

清华大学建筑学院景观学系教授

2024 年 12 月

Foreword 1

Since the dawn of humanity about 6 to 7 million years ago, 99% of our lifetime has been spent in green environments, leading to the formation of a "green gene" within us. The sight of greenery can soothe our minds and contribute to our health, which is the essence of green health therapies, such as horticultural therapy and healthcare programs.

Due to rapid urbanization, habitat fragmentation, severe urban pollution, and reduced biodiversity, there is a widespread prevalence of "urban diseases" and "urban population diseases". "Urban diseases" manifest as traffic congestion, housing shortages, inadequate water supply, energy scarcity, environmental pollution, disorder, an imbalance in material and energy flows, and intensified demand contradictions. "Urban population diseases" are characterized by severe aging trends, an increase in sub-healthy populations, the younger age group with chronic diseases, increased psychological disorders among adolescents, and a noticeable addiction to mobile phones. Research reports suggest that 70% of "urban population diseases" can be prevented and treated through human interventions via green environments. Horticultural therapy and landscape healthcare programs will play a significant role, especially in disease prevention, treatment and recovery from chronic diseases, age-related diseases, and mental health disorders.

In response to current trends and future development needs, the Ministry of Human Resources and Social Security of the People's Republic of China released the newly revised *Occupational Classification Code of the People's Republic of China* (hereinafter referred to as the new version of the "Code") in July 2022. Under Category Four: "Social Life and Living Service Personnel", there is "Subcategory 4-14 (GBM41400) for Health, Sports, and Leisure Service Personnel", which includes seven minor categories of personnel engaged in the auxiliary services such as health consultation, medical clinical practice, rehabilitation therapy, public health, sports health, and wellness leisure. Among these, a new occupation, "Subcategory 4-14-06-01 for Forest and Landscape Healthcare Practitioner", has been added under "Subcategory 4-14-06 (GBM41406) for Healthcare and Leisure Service Personnel". This occupation covers Forest Health Therapist and Garden Health Therapist, and has become officially recognized profession by the state.

Facing the advent of professionalization in landscape healthcare, we must undertake the following tasks to lay a solid foundation for the emergence and development of Garden Health Therapist: ①conduct in-depth scientific research on horticultural therapy and landscape healthcare; ②establish occupational standards for landscape healthcare practitioners; ③strengthen the cultivation of landscape healthcare talents in relevant colleges and universities; ④develop a stringent certification system for landscape healthcare practitioners;

⑤establish high-quality training and internship bases, and improve the certification system; ⑥provide employment opportunities for landscape healthcare practitioners; ⑦implement professional supervision for landscape healthcare practitioners; ⑧further improve the green healthcare practitioner system covering horticultural, landscape, forest, and agricultural health care.

A Bilingual Textbook on Healing Garden Design, is an important component of the aforementioned work. Presented in both Chinese and English, it elaborates on the theoretical underpinnings and practical applications of healing gardens, including the concepts, historical development, theoretical foundations, design principles, as well as applications of different functional healing gardens. In addition, it also attaches great importance to the origins of healing gardens within Chinese garden culture. The integration of Eastern and Western perspectives in the textbook can not only cultivate students' international vision, but also demonstrate Chinese characteristics and cultural confidence to a large extent. In terms of ideology, scientific nature and advancement, the textbook can meet the teaching requirements. Therefore, it will surely play a significant role in promoting the development of landscape wellness and horticultural therapy in our country.

Professor Li Shuhua
Department of Landscape Architecture,
School of Architecture, Tsinghua University
December 2024

序 2

　　历经数百万年的繁衍生息，人类逐渐发现大自然存在某种特殊环境，除了常见的山清水秀、鸟语花香，还具有修复自身病灶、恢复健康的功效。面对这一大自然的馈赠，时至今日，在边远民族区域，仍然可见人们深入这种原始天然的自然环境，借以获得自然康复的功效。在这一久远广阔的人—境互动演进中，"移天缩地""小中见大""巧于因借"等，将这种自然环境"有意为之"地设计建造，康复花园的出现也就水到渠成了。

　　现代康复花园的设计建造将植物资源的多样性、植物对环境的生态贡献以及对人类的健康功效体现在当代人居环境——园林绿地的建设中，充分体现出人类借助风景园林追求对自然环境感、知、应最佳状态的哲学思想和营造手法，既是实现风景园林能量接受、信息传递、时空应对的"三元耦合"和人与天地"二因互动"的典范，也是当代最能唤醒初心、回归人性、令人流连忘返的仙境再现。

　　与其源远流长的实践应用相比，康复花园的理论研究还只是近现代的事，从设计建造到运营管理的现代科学专业化实践也积累不多，而专业人才的培养更是刚刚起步。在为数不多的关于康复花园的专著教材中，本书集现代康复花园设计的理论研究、实践应用、专业教育"三位一体"，当属首创且特色鲜明。同时，基于起步阶段，主要面向本科专业学生的普及入门教育，全书编写内容基础性强并通俗易懂，为日后的各方面提升深化和发展奠定了基础。毫无疑问，随着人类社会文明水平的提升，该书的价值将逐渐展现，与日俱增。

澳门科技大学人文艺术学院建筑学与设计学教授

2024 年 10 月于上海

Foreword 2

Over millions of years of reproduction and survival, human beings have gradually discovered that there are certain special environments in nature, which bestow on them not only the beauty of green mountains and clear waters, the fragrance of flowers and the melodies of birds, but also the power to heal ailments and restore health. Faced with this gift from nature, even today, in remote areas inhabited by ethnic minorities, it can still be seen that people venture into these primitive natural environment to obtain the benefits of natural rehabilitation. In this far-reaching evolution of human-environment interaction, the "intention-based" design and construction concepts of the natural environment, such as "shrinking the world", "seeing the big in the small", and "skillfully borrowing", have made the emergence of healing gardens a natural progression.

In the design and construction of modern healing gardens, the diversity of plant resources, the ecological contributions of plants to the environment and their health benefits to humans are incorporated into the development of green spaces in contemporary human settlements and landscape environments. This fully embodies the philosophical thought and crafting techniques of humans seeking the optimal state of perception, cognition, and response to the natural environment through landscape architecture. These gardens serve as a model for the "ternary coupling" of energy reception, information transmission, and space-time response in landscape architecture, as well as the "two-way interaction" between humans and nature. They are also the most enchanting contemporary paradises that can awaken our original intentions, allow us to return to humanity, and make us linger on.

Compared to its long history of practical application, the theoretical research on healing gardens is a more recent phenomenon. There is still a limited accumulation of contemporary specialized scientific practices from design and construction to operation and management, and the cultivation of professional talents has just started. Among the few monographs and textbooks on healing gardens, this book pioneers "the distinctive trinity" of the theoretical research, practical application, and professional education of modern healing garden design. Meanwhile, based on the initial stage and primarily aimed at undergraduate students for introductory education, this book is fundamental and easy to understand, laying a foundation for future advancements and developments in all aspects. Undoubtedly, as the level of human civilization progresses, the value of this book will gradually reveal itself and increase over time.

Professor Liu Binyi
Department of Design and Architecture,
Faculty of Humanities and Arts, Macau University of Science and Technology
Shanghai, October 2024

前言

"何以解忧，唯有园林"——这是我国著名建筑学家、园林艺术家陈从周先生（1918—2000年）借当年曹操之语，以表达园林对人身心调养之意。无独有偶，美国著名园艺学家路赛·布尔斑克（Luther Burbank，1849—1926年）也曾说过一句名言："Flowers always make people better, happier, and more powerful; they are sunshine, food and medicine for the soul."（花卉总能让人更美好、更愉悦、更有力量；它们是心灵的阳光、食粮，更是良药）。名噪一时的畅销书《醒来的森林》的作者、美国知名自然主义作家约翰·巴洛弗斯（John Burroughs，1837—1921年）说道："I go to nature to be soothed and healed and to have my senses put in order."（我投身于大自然中去，让自己减轻痛苦、获得疗愈，让我所有的感知更加平衡有序）。这些名言都表明，大自然具有抚慰感知、疗愈身心的功能。在前人心目中，花卉、园林能让人获得身心平衡和幸福感，相信这是所有热爱自然之人共有的体验，而每个人心里也大抵装着一个花园梦。

在都市人群与大自然渐行渐远的现代社会，工作的压力、快节奏的生活都让人备感倦怠。回归自然似乎成了奢侈的享受。人们利用休假日成群结队奔赴大自然，节假日景点爆棚、高速路拥堵等现象已成常态。是时候让自然回归城市了！新近推出的第四代建筑——空中庭院让我们看到了亲近自然的曙光。在城市里重建自然、让园林重回"呵护人类身心健康"的初衷，是我们现代风景园林人新时期的历史使命和责任。

自古就有"养花植草，修身养性"之说的中华民族，历来有植树种花的习俗，也深谙植物对人的保健功能以及空间配置之道。不管是祖国瑰宝《中草药学》，抑或留给我们的古典园林，无一不是"天人合一"的自然养生之作，展现出中国古人利用花园与植物获得身心健康的智慧；古埃及也有文字记载用花园疗愈皇室成员疾病的案例。回顾中外历史，我们不难看出，自人类鸿蒙之初，人们对自身栖息之所的需求首先是要从中得到身体的舒适，进一步要获得心灵的释然。所以，具备疗愈与康复的功能，便成为营建人居环境的重要目的之一。

然而，工业文明的到来，在把人类从繁重的体力劳作中解放的同时，也极大地消减了自然作为人类疗愈场所的重要性。加之近百年来，现代医疗技术的飞速发展也让人们养成了"重治疗、轻关怀"的健康维系方式。"治未病、强身心"被人淡忘，人们对自己的身体多了放纵、少了约束，直到万不得已，生了病才去治疗。这样的模式，不但耗费了大量的社会资源，人们的生活品质与工作效率也深受影响。所幸，为了纠正这种违背自然的生活方式、提升现代社会健康照护的效率，东西方不少学者又开始重新回到中医以及西方的前现代医学等亲自然的疗愈体系中，去寻找古人的康复智慧，园艺康健（Horticultural Therapy）与康复花园（Healing Garden）等概念也因此应运而生。相比古人的文献，这些概念虽然是新的，但实质内涵却是源于古代文明的传承，这不能不说是人类文明在健康照护领域的波浪式前进、螺旋式上升。关于园艺康健的概念，在此做一简要说明：该概念源自对英文 Horticultural Therapy 的翻译。前些年"therapy"被译为"疗法""治疗""福祉""疗愈"等不同词汇。然而 therapy 和 treatment 不同，前者是非医疗的复健方法。为避免和医疗相关概念混淆，本教材采纳程宗明教授的翻译，将其译为"康健"，这样更符合中国人对健康照护的理解。"康"意为从不良状态到正常状态的恢复，"健"意为正常状态到更强健的发展。

如今，全球文化正在调转方向，朝着人对自然的包容性、依赖性理念发展；城市历史学家山姆·巴斯·沃纳（Sam Bath Warna）将这种趋势称为"对共同智慧的重新探索"。社会生态学家斯蒂芬·科勒特（Stephen Collet）则认为，这种回归自然的趋势是我们与生俱来的权利；世界著名生物学家威尔逊（E. O. Wilson）将其描述为人类热爱生命的天性，是"与其他生命形式紧密联系的强烈欲望"。显而易见，与自然保持联系，对健康和幸福而言是十分必要的，包括减轻压力、缓解糟糕的情绪、得到锻炼和增进社交联系、少用药物的情况下加快康复的速度等。同时，这种转变也需要我们作为人居环境营造者——风景园林师敏锐的、适宜的、创新的设计反应。通过设计，将人类健康与环境健康联系起来，从个体到集体，让社会、城市和世界更富生气、更加平衡。这正符合了康复花园诞生的初衷。

党的二十大报告提出，要深刻把握生态文明建设这个关乎中华民族永续发展的根本大计，扎实推动绿色发展，促进人与自然和谐共生，共同建设美丽中国；着眼全面建设社会主义现代化国家的目标任务。对增进民生福祉、提高人民生活品质作出重要部署时强调：必须坚持在发展中保障和改善民生，鼓励共同奋斗创造美好生活，不断实现人民对美好生活的向往。康复花园的设计建造和运用，正是风景园林基于景观感应新理论（刘滨谊，2022），立足文化自信，促进人与自然和谐共生，共同建设美丽中国、健康中国，满足人民美好生活愿望，实现中国式现代化的一条重要路径。

康复花园一词，产生于 20 世纪 90 年代的美国，康复花园是实施园艺康健的室外场所。美国园艺康健协会 2012 年修订的相关术语解释文件中，将其解释为："特指以植物占主导的有恢复身心特定功能的环境，包括绿色植物、花卉、水和其他自然元素。康复花园服务于所有人，除了日常生活所需，康复花园还与医院和其他医疗保健设施相关联，旨在为体验者和游客及其工作人员提供静思和喘息的场所。"康复花园可以进一步划分为特定的花园类型，包括治疗花园、冥想花园、记忆花园和疗养花园等。作为与园艺康健配套的环境，与普通花园相比，康复花园更强调设计的循证基础，即满足特殊的功能需求、更加以人为本的需求。而园艺康健是通过训练有素的园艺康健师引领参与者实施园艺操作活动或者与植物相关联的各种活动以达到恢复身体机能、恢复心理健康以及增进社会适应能力等目标的方法。康复花园与园艺康健是一对不宜分割的概念，没有园艺康健的康复花园是缺乏内涵的，难以使人达到真正康复的有效目标；而没有康复花园的园艺康健则少了环境的依托，略显单调。因此，通过照护植物的过程去实现人们自身身心健康的这个模式中，康复花园与园艺康健成了不可或缺、互为支撑的硬件与软件。

关于康复花园的词义，在美国有 Healing Garden、Restorative Garden、Horticultural Therapy（H. T.）Garden 等说法，日本引用 H. T. Garden 较多，李树华于 2000 年将 H. T. Garden 翻译为中文，称其为园艺疗法庭院（李树华，2000）。相比于 restorative 一词，healing 的含义更广泛，后来美国园艺康健协会将以上名称统一为 Healing Garden，该词义的中文翻译，目前在中国有多种版本。俞孔坚等翻译的克莱尔·库珀·马库斯等编著的《人性场所——城市开放空间设计导则》一书，将 Healing Garden 译为医疗花园，将 Restorative Garden 译为康复花园（Marcus，2001），后金荷仙等人翻译的克莱尔·库珀·马库斯 *Healing Garden in Hospital* 一文，直接将 Healing Garden 译为康复花园（Marcus，2009）；杨欢等人则翻译为康健花园（杨欢，2009）；之后亦出现疗愈花园、治疗景观等翻译。本书在编译过程中亦做了反复推敲，结合近年来"康复花园"这一概念在受众中的普及率，最终采用"康复花园"的翻译。汉语中，"康复"一词在南朝裴松之为《三国志》做的注当中就已出现："康复社稷，岂曰天助，抑亦人谋也"；1948 年世界卫生组织（WHO）宪章中，对"健康"一词的定义为"健康不仅为疾病或赢弱之消除，而系体格、精神与社会之完全健康状态"。因此 WHO 将"康复"定义为："泛指综合地、协调地应用医学的、教育的、社会的、职业的各种措施，使病、伤、残者（包括先天性残疾者）已经丧失的功能尽快地、最大可能地得到恢复和重建，使他们在体格上、精神上、社会上和经济上的能力得到尽可能的恢复，重新走向生活、走向工作、走向社会。"可见，康复不仅针对疾病，而是着眼于整个人体，从生理上、心理上、社会上及经济能力上进行全面恢复。

所以，康复花园的功能在于人们能够尽可能地从中获得体格、精神、社会和经济能力的恢复。特别是结合园艺康健的实施，针对那些需恢复社交以及经济能力的人们，除了身心的康复，他们还能学习园艺的技术，掌握一技之长重返社会。

然而康复花园的设计在我国刚刚起步，目前尚无同类代表性教材出版，且有关康复花园设计的文献多为英文。鉴于此，我们编写了本教材，以推进我国康复花园设计的教学实践研究与国际接轨；同时，中英双语教程更适合读者学习理解该领域的英文专业术语，有助于提升外文文献的阅读，及时了解该领域的国际研究动态。本教材围绕康复花园的内涵和外延展开阐述，从其概念、发展沿革、理论基础、研究动态到不同功能型康复花园的设计应用等均用中文和英文进行论述和讲解。本教材可以作为风景园林专业相关教师、学生，园林工程师或其他从业人员的阅读书籍和参考资料。

本教材由郑丽、陈利平任主编，编写分工如下：前言由郑丽编写，第 1 章由吴祥艳、郑丽、陈志欣编写；第 2 章由郑丽编写；第 3 章由袁晓梅、郑丽编写；第 4 章由孙旻恺、郑丽编写；第 5 章由吴祥艳、郑丽、乌云巴根、陈飞、陈立人编写；第 6~10 章、附录由吴祥艳、郑丽、袁惠燕、付晓渝、李茜玲、王胜、陈志欣、李房英、孙志勇、包晓鹏、李燕、黄印星、谭凯欣编写。翻译人员及分工：前言与第 1 章由陈友华翻译；第 2 章由吴广平翻译；第 3 章由石杉杉翻译；第 4 章由王晓蕾翻译；第 5

章由赵萍翻译；第6~10章、附录由石杉杉、陈志荣、白晶翻译。陈利平、张吟松负责全书译文统审。美国园艺康健协会（AHTA）学术研究组前组长、纽约大学郎格尼医学中心临床副教授 Matthew Wichrowski 对全书英文进行了校审，清华大学建筑学院李树华教授、上海交通大学设计学院姚雷教授对全书中文进行了审定。

当然，康复花园并非局限于本教材所编写类型，教材涉及类型为目前该领域实践运用较为常见者。我们需要理解的是，康复花园是一种引领和改变行为模式的设计，通过设计让人们的身心获得调适，从而保障健康。

本教材在编写过程中，得到了中国林业出版社以及苏州大学教务处的大力支持，特别感谢苏州大学曹林娣教授对中国传统园林文化的指导，尤其感谢李树华教授、刘滨谊教授为本书赐序；同时本书的编写工作也得到了云南农业大学盛军教授，华南理工大学袁晓梅教授，福建农林大学吴少华教授，原重庆市风景园林科学研究院院长先旭东研究员，南京农业大学陈素梅教授，西南大学李先源教授，苏州园科生态集团董事长毛安元研究员的鼎力支持。在此深表感谢！同时也感谢徐虹、朱莉红、黄宇欣、王彦人、王震、王一舒、莫馨彦、朱同、纪霖、翁璐、陈志颖、闫锦、白彬奕、郑梦丽、赵柏婷等人在编校过程中给予的支持与帮助。最后，再次感谢主审李树华和姚雷两位教授在百忙之中给予的指导。

由于编者水平有限，书中错误在所难免，敬请读者批评指正！

编 者

2024 年 6 月

Preface

"It is landscape art that can relieve our melancholy" —Mr. Chen Congzhou (1918— 2000), a well-known architect and landscape artist in China, quoted what Cao Cao once said, to express the soothing effect of landscape art on human body and mind. Coincidentally, Mr. Luther Burbank (1849—1926), a famous American horticulturalist, also believed that "flowers always make people better, happier, and more powerful; they are sunshine, food and medicine for the soul." John Burroughs (1837—1921), the well-known American naturalistic writer of the best-selling book entitled *Wake-Robin*, said that "I go to nature to be soothed and healed and to have my senses put in order." These quotations indicate that nature is endowed with the functions of soothing human perception and healing human body and mind. From their perspectives, flowers and landscape art enable people to enjoy balance in body and soul, and sense of happiness. Convincingly, it is the shared experience for people who love nature. And there is no doubt that a dream of garden is deeply rooted in our mind.

In the modern society where urban people are moving further and further away from nature, and where people are exhausted under the pressure of work and the fast pace of life, returning to nature seems to be a luxury. Fleeing the concrete jungles in flocks during the holidays, people rush into the realm of nature, which leads to the common phenomena, such as the bursting holiday tourist attractions and highway congestion. It is time to restore nature to urban areas! The dawn of getting closer to nature breaks as the housing style of the fourth-generation architecture's air courtyard is newly launched. It is the historical mission and responsibility in the new era for our modern landscape gardeners to rebuild nature in the city, and make landscape art revert to its original intention of "caring for the physical and mental health of mankind".

Under the firm belief that "planting trees and flowers is healthy and beneficial to one's self-cultivation", the Chinese have had the classical custom of planting trees and flowers since the ancient times. They are quite familiar with plants' healthy functions and the arts of space allocation. No matter whether it is our treasure of *Chinese Herbology* or the heritage of our classical gardens, all of them are the masterpieces of natural hygiene, demonstrating the concept of "Harmony between Man and Nature" and the ancient Chinese wisdom of maintaining physically and mentally healthy by means of gardens and plants. There are recorded cases of curing the royal members of their diseases in ancient Egypt. In retrospect, it is easy for us to find that people's demand for their dwellings has been to first obtain physical comfort and then to relax their mind since the dawn of human history. Therefore, having the function of healing and rehabilitation becomes one of the primary aims in constructing habitable environment.

The arrival of industrial civilization liberates people from hard work. However, it also greatly reduces the importance of nature as a place for human rehabilitation. For the past century, the rapid development of modern medical technology has seen the establishment of health maintenance style that "medical treatment outweighs solicitude". It is gradually forgotten to "prevent diseases and strengthen one's body and mind". There is more physical indulgence than restraint and people seldom go to hospital unless they have to. This style not only consumes a large amount of social resources, but also greatly affects people's life quality and work efficiency. Fortunately, to correct such a lifestyle against the law of nature and improve the health care efficiency of modern society, quite a few scholars at home and abroad return again to the traditional Chinese medicine, pre-modern western medicine and other pro-nature healing systems in the quest of the ancient wisdoms of recovery. Hence the birth of such concepts as horticultural therapy and healing garden. Compared with the ancient literature, the fresh concepts originate from the ancient civilizations in essence, which shows the tendency of human civilization's wave-like advance and spiral escalation in the field of health care. Here is a brief explanation for the translation of "Horticultural Therapy" into Chinese. In the past, "therapy" was translated into "Liao Fa" "Zhi Liao" "Fu Zhi" "Liao Yu" and so on. However, different from "treatment", "therapy" refers to non-medical methods of rehabilitation. In this book, to avoid confusing with relevant medical concepts, Professor Cheng Zongming's translation of "therapy" is adopted: "Kang Jian" is in better line with the Chinese people's understanding of health care. In Chinese, "Kang" means the recovery from the poor health state into the normal. "Jian" refers to the development of the normal state of body into the stronger.

Global cultures are reorienting themselves towards human tolerance of nature and dependence on nature. City historian Sam Bath Warna addressed this kind of tendency as "the re-exploration of the shared wisdoms". Social ecologist Stephen Collet thought that the tendency of returning to nature is the power that mankind are born with. E. O. Wilson, the world-known ecologist, described it as human nature of love for life and "the strong desire for the close connection with other life forms". Obviously, it is quite necessary for human welfare to keep connected with nature, for example releasing pressure and bad mood, having access to adequate physical exercises and social connection, accelerating rehabilitation with reduced pills. Meanwhile, this transformation also needs our landscape architects' acute response to the suitable and innovative designs as the builders of habitable environment. The original intention of the birth of healing gardens is just to connect human health with healthy environment, from individuals to the collective, and to make our cities, society and the world more animate and more balanced through these designs.

As the Report of the 20th National Congress of the Communist Party of China points out, we should fully understand that ecological civilization construction is a fundamental task for the sustainable development of the Chinese nation. We should steadily promote green development and the harmony between man and nature, and work together to build a beautiful China. And we should focus on the goals and tasks of building a modern socialist country in an all-round way. In drawing up important plans for improving public wellbeing and the quality of life, the 20th National Congress emphasizes that we must persist on ensuring and improving public wellbeing in the course of development. And we should encourage national concerted efforts to build a better life and to constantly realize the national aspiration for a better life. The design, construction and application of healing gardens are just based on the new theory of landscape perception (Liu Binyi, 2023). An important path to realize Chinese modernization is to be based on cultural confidence, to promote harmonious coexistence between man and nature, to jointly build a beautiful and healthy China, and to meet the national desire for a better life.

The term, Healing Garden, originated in the United States in 1990s. It refers to the outdoor places for horticultural therapy. According to the 2012 revision of garden glossary by the American Horticultural Therapy Association (AHTA), it is defined as "the plant-oriented environment with specific rehabilitation functions, including green plants, flowers, water and other natural elements. Healing garden is open to all of people. In addition to meeting the needs of daily life, it is mostly associated with hospitals and other medical health care facilities, aimed at providing users, visitors and its staff with a place for contemplation and relaxation". Healing garden can be further divided into specific garden types such as therapy garden, meditation garden, memory garden and recuperation garden. Acting as the matching environment with horticultural therapy, healing garden focuses more on the evidence-based design, compared with ordinary ones. That is, its design should satisfy the needs for specific functions and be more people-oriented. Horticultural therapy is aimed at the recovery of physical functions, mental health and social adaptation ability by participating in horticultural practical activities and other various plant-related activities under the guidance of well-trained horticultural therapists. Healing garden and horticultural therapy is a pair of undivided concepts. Healing garden without horticultural therapy is empty and unable to achieve the effective target of rehabilitation, while horticultural therapy without healing garden is a bit monotonous for lack of the environmental support. Therefore, healing garden and horticultural therapy have become indispensable and mutually supporting hardware and software realizing the mode of human physical and mental health by way of taking care of plants.

There are various expressions for rehabilitation garden. It is addressed as Healing Garden, Restorative Garden, and Horticultural Therapy (H. T.) Garden in America and so on and mostly as H. T. Garden in Japan. Mr. Li Shuhua translated it into Chinese, known as Horticultural Therapy Yard (Li Shuhua, 2000). Compared with the word "restorative", "healing" has more profound connotations. Afterwards, the American Horticultural Therapy Association unified all the expressions into Healing Garden. There are many versions of Chinese translation of the term *Healing Garden* in China. It first occurred in *People Places: Design Guidelines for Urban Open Space*, edited by Clare Cooper Marcus et al and translated by Yu Kongjian et al. Healing Garden was translated as "Yiliaohuayuan", and Restorative

Gardens as "Kangfuhuayuan" (Kelaier, 2001). And then, in the book *Healing Garden in Hospital* written by Clair Cooper Marcus, Jin Hexian et. all translated it as "Kangfuhuayuan" (Kelaier, 2009). Yang Huan translated it as "Kangfuhuayuan" (Yang Huan, 2009). Afterwards, there appeared some other versions of Chinese translation like "Liaoyuhuayuan" and "Zhiliaojingguan". This book adopts the term "Healing Garden" (Kangfuhuayuan) after repeated consideration in the process of editing and translation. In China, as early as in the Southern Dynasties, the term "healing" first occurred in Mr. Pei Songzhi's annotations for *The Records of the Three Kingdoms*, which said "rehabilitation in a nation cannot solely depend on god, but also on the effort of people themselves". The Charter of the World Health Organization (WHO) in 1948 defined health as "not merely the absence of disease or infirmity, but a state of completely physical, mental and social health", and defined rehabilitation as "trying one's best to recover and reconstruct the lost functions of the sick, the injured and the handicapped as soon as possible by comprehensively and harmoniously applying a variety of medical, educational, social and professional measures, and helping them recover their physical, mental, social and economic abilities as much as possible, so that they can return to their work, life and society." Rehabilitation is aimed not only on illness, but also on the complete recovery of an individual in terms of one's physiology, psychology, social and economic abilities.

So the function of healing garden lies in the rehabilitation of body, soul, social and economic abilities as much as possible. Especially in combination with horticultural therapy, healing garden aims at those who are in need of the recovery of their social and economic abilities. Besides their physical and mental recovery, healing garden can also help them grasp horticultural skills to return to the society.

However, since healing garden design just sets foot in China, not any similar representative books in this field have been published for colleges and universities at present. Moreover, literature on healing garden design is mostly written in English. Hence the compilation of this book to promote the teaching, practicing and researching of healing garden design in China in line with international standards. Meanwhile, contributing to the comprehension of English literature and the knowledge of international research developments, bilingual books are more suitable for learners to have an idea of English technical terms in this field. Focusing on the connotation and extension of healing garden, this book discusses and elaborates both in Chinese and in English on the concepts, the history of development, theoretical basis, research developments and the design and application of healing gardens with various functions and the like. This book can also act as a reference book for teachers, landscape architecture majors, landscape engineers and other practitioners.

Professor Zheng Li and Professor Chen Liping are the editors in chief of this book. The compilation team are as follows: Preface, compiled by Zheng Li; Chapter 1, by Wu Xiangyan, Zheng Li and Chen Zhixin; Chapter 2, by Zheng Li; Chapter 3, by Yuan Xiaomei and Zheng Li; Chapter 4, by Sun Minkai and Zheng Li; Chapter 5, by Wu

Xiangyan, Zheng Li, Wu Yun Ba Gen, Chen Fei and Chen Liren; Chapter 6 to 10, and Appendixes by Wu Xiangyan, Zheng Li, Yuan Huiyan, Fu Xiaoyu, Li Qianling, Wang Sheng, Chen Zhixin, Li Fangying, Sun Zhiyong, Bao Xiaopeng, Li Yan, Huang Yinxing and Tan Kaixin. The translation team are as follows: Preface and Chapter 1, translated by Chen Youhua; Chapter 2, by Wu Guangping; Chapter 3, by Shi Shanshan; Chapter 4, by Wang Xiaolei; Chapter 5, by Zhao Ping; Chapter 6~10 and Appendixes, by Shi Shanshan, Chen Zhirong and Bai Jing. The translation parts are reviewed by Professor Chen Liping, Professor Zhang Yinsong of Yunnan Agricultural University, and Mr. Matthew Wichrowski, Associate Professor of Clinical Practice at Langone Medical Center, New York University, USA. The Chinese parts are reviewed by Professor Li Shuhua from Architecture School of Tsinghua University and Professor Yao Lei from School of Design at Shanghai Jiao Tong University.

Of course, healing gardens are not merely limited to the types compiled in this book, but these types represent the comparatively common ones in this field. What we need to understand is that healing garden is a design to lead and change our behavior patterns, through which human body and mind are adjusted to ensure health.

In the process of compilation, this book obtained the strong support from China Forestry Publishing House and the Office of Academic Affairs at Soochow University, and the valuable guidance of Professor Cao Lindi of Soochow University on the Chinese traditional garden culture. In particular, our gratitude should go to Professor Li Shuhua and Professor Liu Binyi for contributing forewords to this book. This book also obtained the strong support from Professor Sheng Jun of Yunnan Agricultural University, Professor Yuan Xiaomei of South China University of Technology, Professor Wu Shaohua of Fujian Agriculture and Forestry University, Research Fellow Xian Xudong, the former dean of Chongqing Garden Research Institute, Professor Chen Sumei of Nanjing Agricultural University, Professor Li Xianyuan of Xinan University and Research Fellow Mao Anyuan of Suzhou Yuanke Ecological Group. Much appreciation owes to all of them! During the editing and proofreading process, their support and assistance are also greatly appreciated: Xu Hong, Zhu Lihong, Huang Yuxin, Wang Yanru, Wang Zhen, Wang Yishu, Mo Xinyan, Zhu Tong, Ji Lin, Weng Lu, Chen Zhiying, Yan Jin, Bai Binyi, Zheng Mengli, and Zhao Baiting. Finally, our gratitude should once again go to the reviewers, Professor Li Shuhua and Professor Yao Lei, for their guidance in their hectic schedules.

Given the limitations of the editors, it is inevitable that mistakes may occur in the book. Sincerely welcome criticism and corrections!

The editors
June 2024

目　录

Contents

第1章

康复花园起源和发展特征

Chapter 1 The Origin and Development of Healing Garden

狭义的康复花园指与医养环境相结合，针对某些特殊的功能需求而设计，以满足康复训练条件、增强患者战胜病痛信心的花园；广义的康复花园泛指可与园艺康健配套、满足一定使用功能的花园，是实施园艺康健的场所，可视为园艺康健的硬件支撑系统之一，可以进一步划分为特定的花园类型，包括治疗花园、冥想花园、记忆花园和疗养花园等（郑丽，2020）。康复花园是人类对自身居住环境从绿化、美化，到生态化，再到健康化的逐级追求中，所达到的较精细的环境表现形式，可谓是当今大健康背景下，园林环境与健康之间衍生的一种新型关系。随着健康照护模式由单一的生物医学模式向"生物—心理—社会"综合模式的转变（梁友信，1987），园林环境对健康的促进作用更体现出其重要性。生物—心理—社会医学模式（biopsychosocial model）最早是由美国精神病学家乔治·恩格尔（George Engel）在1977年提出的。恩格尔批评了当时主流的生物医学模式，认为它忽视了心理社会因素对健康的影响。生物—心理—社会医学模式强调，在考虑个体的健康和疾病时，不仅要考虑生物学因素，还要考虑心理、社会和环境因素的综合影响。这些因素分别是：①环境因素——包括物理、化学、社会、经济和文化等因素；②行为及生活方式——包括营养、风俗习惯、嗜好、吸烟、酗酒、交通事故及精神紧张等；③卫生服务——包括医疗卫生设施及其服务制度；④生物遗传因素。四个因素中，环境因素最为重要，生活方式次之，卫生服务居第三位，最后是生物遗传因素（George，1977）。

In a narrow sense, healing garden is designed in combination with the medical environment for some specific functional needs to satisfy the rehabilitation training conditions and strengthen patients' confidence in conquering illnesses and pains. In a broad sense, healing garden refers to the garden that can be compatible with horticultural therapy and meet certain functional needs. It is the place for the implementation of horticultural therapy and can be regarded as one of its hardware support systems. It can be further divided into such specific types of garden as therapy garden, meditation garden, memory garden and recuperation garden (Zheng Li, 2020). In the pursuit of different levels of human living environment, from the green environment to the beautified environment, to the ecological environment, and then to the healthy environment, healing garden is a comparatively refined form of environment, and also a new relationship between garden environment and human health in the context of modern big health. In the transition of health care mode from a single "biomedical model" to a comprehensive "biopsychosocial model" (Liang Youxin, 1987), garden environment is becoming increasingly important in promoting health. Biopsychosocial Model was first put forward by George Engel, an American psychiatrist, in 1977. He criticized the mainstream biomedical model of the time, arguing that it ignored the impact of psychosocial factors on health. Biopsychosocial Model emphasizes not only the influence of biological factors, but also the combined influence of psychological, social and environmental ones as far as individual health and disease are concerned. To be specific, these factors are as follows: ①environmental factors, including physical, chemical, social, economic and cultural factors; ②behavior and lifestyle, including nutrition, customs, hobbies, smoking, alcoholism, traffic accidents and mental strain; ③health services, including medical and health facilities and service systems; ④biological genetic factors. Among the four factors, environmental factors are the most important, lifestyle is the second, health services are the third, and the biological genetic factors are the last (George, 1977).

1.1　从城市绿地到康复花园
1.1　From urban green space to healing garden

1.1.1　城市绿地
1.1.1　Urban green space

与乡村景观相比，以公园为代表的城市绿地为城镇较为密集的建筑提供了自然元素，为居民提供消遣游憩的场所，同时起到美化环境、改善生态的作用。现有城市绿地、公园是设计者从美学和生态学角度出发，依靠理性的经验和感性的创作营造出的具有一定功能和用途的公共绿地，面向所有城市人群，具备一定的功能分区，但功能用途相对单一；绿地内的植物选择能满足季相变化，常选用乡土树种；评价主要包括生态、艺术、经济、功能等方面。

In contrast with rural landscape, the emergence of urban green space, with parks as its representatives, provides the comparatively dense urban buildings with natural elements and recreational places for the public. Meanwhile, it plays the role of beautifying environments and improving ecology. From the perspectives of aesthetics and ecology, the current urban green space and gardens are the public green space with certain functions, which are created by designers on the basis of rational experience and perceptual creation. They are open to the urban population with a certain function zones, but with relatively limited uses. The choice of plants in the green space varies with seasons and native species of plants are preferred options. The construction effect of urban green space is mainly assessed from the following aspects such as ecology, art, economy and function.

1.1.2 康复花园产生的时代背景
1.1.2　The historical background of the birth of healing garden

自从 20 世纪 60 年代开始，蕾切尔·卡森的《寂静的春天》把人们从工业时代富足的梦想中唤醒，随后加勒特·哈丁在《哈丁悲剧》中进一步揭示了人类本性中的贪婪和资本主义经济导致的资源枯竭现象……人类生存的危机开始逐步把设计师们从对美与形式的陶醉中引向对人与自然关系的关注。于是，推崇保护自然、节约资源的生态思想应运而生。在此背景下，英国景观设计师和生态规划的倡导者麦克哈格在《设计结合自然》一书中提出其景观思想和方法，产生了更为广泛意义上的生态设计，包括景观和城市设计的生态规划（翟俊，2018）。景观设计师也逐渐将生态思想引入景观概念和营造中，形成具有科学内涵的生态景观。

Since the 1960s, Rachel Carson's *Silent Spring* had awakened people from the dream of prosperity in the industrial age. Afterwards, in *The Tragedy of the Commons*, Garrett Hardin further revealed the greed of human nature and resource exhaustion caused by the capitalist economy. The crisis of human survival gradually led designers to the concern with the relationship between mankind and nature, away from the intoxication with beauty and forms. Hence the birth of the ecological thought of advocating the conservation of nature and resources. In this context, Ian McHarg, the British landscape architect and advocate of ecological planning, put forward the landscape theories and methods in *Design with Nature*, which led to ecological designs in a broader sense, including ecological planning in landscape and urban design (Zhai Jun, 2018). Landscape designers also gradually introduced the ecological thought into the landscape concepts and construction. Thus ecological landscape with scientific connotation came into being.

虽然生态景观目前仍是现代景观设计的主流思想，但健康亦不可避免地成为全人类更为关注的话题。因此，以人为本的增进健康的环境设计应运而生，这就是康复花园景观设计。在康复花园景观设计过程中，不仅要实现环境生态理念，也要从使用者自身出发，在保护自然的同时，更要兼顾良好的观赏性、功能性和参与性，最终实现使用者身心健康照护的目标。

Although ecological landscape has still been the mainstream of modern landscape design at present, health also inevitably becomes the more concerned topic for human beings.

Hence the birth of the people-oriented design to improve health and environment: the design of healing garden landscape. In the process of designing healing garden landscape, it should be ensured to realize the ecological thought of environment. In the meantime, from the standpoint of the users themselves, what should be taken into account is ornament, functions and users' participation, with its ultimate goal of enabling the users to be physically and mentally healthy.

1.1.3　康复花园的特点
1.1.3　Features of healing garden

健康城市理论认为，健康不仅取决于医学因素，更取决于广泛的社会、环境等因素。健康不再是单纯的医疗方面的问题，也是生理、心理以及社会、环境各个方面的综合问题，这使得兼顾健康理念的景观设计有了更广的范围和更深刻的内容。

According to the Healthy Cities theory, health is not only related to the medical factors, but also more related to a wide range of social and environmental factors. In addition, health is not simply a medical problem, but a comprehensive problem concerned with physiology, psychology, society and environment, which leads the landscape designs with the concept of health to a wider scope and a more profound content.

始于 20 世纪后期的康复花园，其前身可以追溯到东方的寺院园林以及欧洲的修道院园林。康复花园包含了现有城市绿地、公园的构成要素和特征，但又不局限于景观和生态层面。康复花园面向特定使用人群，准确定位他们的需求和心理特征，充分挖掘花园的康复特质，通过调动使用者的感官体验，以期实现特殊疗愈作用。康复花园的康复功效主要包括三个层次：缓解身体不适、减轻心理压力和增强幸福感。因此，康复花园功能分区较之于城市普通绿地更为详细而深入，能最大化满足使用者需求，充分体现人文关怀；花园内的植物在满足美学要求的前提下，多采用具有疗愈功效的植物，如药用植物、芳香植物、净化空气类植物等，并充分挖掘这类植物的功能属性；花园的设计营造过程贯穿了"寻找证据—运用证据—总结成果"的循证研究，设计师需要多次评估康复景观的治疗价值，这是康复花园不同于现有城市公园绿地的一大特征。营造效果评价主要包括：感知环境、无障碍环境、活动空间、人体功效环境、管理和维护、花园容纳量等要素。康复花园常与园艺康健相结合，设计时需预留出园艺康健区。近年来，一些有特色的康复花园，如"冥想花园""五感园""香花诊室""温泉浴""森林浴"等开始涌现。

Born in the late 20th century, healing garden can be traced back to the monastery gardens in Europe and the Oriental temple gardens. Healing garden contains the components and features of the existing urban green space and parks, but is not limited to its landscape and ecological level. Targeted at specific users and accurately oriented to their needs and psychological characteristics, healing garden can hopefully achieve its special healing effect at three levels through fully exploring its rehabilitation characteristics and arousing the users' sensory experience: relieving physical discomfort, reducing psychological pressure, and enhancing sense of happiness. Therefore, in contrast with the ordinary urban green space, healing garden has more detailed and complicated division of function zones to maximize the users' satisfaction of needs and fully embody humanistic concern. On the premise of satisfying the aesthetic requirements, healing garden is more likely to choose plants with healing effect, such as medicinal plants, aromatic plants and air-purifying plants. Moreover, their functions and traits will be fully explored. Different from the existing urban parks and green space, a major feature of healing garden is that its design and construction follow the evidence-based research of "evidence searching, evidence application and achievement review". Designers need evaluate the healing value of healing garden landscape time and again. As far as the construction effect is concerned, its indicators mainly cover such elements as perception environment, barrier-free environment, activity space, ergonomics environment, management and maintenance, and garden capacity. Healing garden is often combined with horticultural therapy, and horticultural therapy areas need to be reserved in its design. In recent years, some healing gardens with different features have begun to spring up, for example, "meditation garden", "five senses garden", "fragrant flower therapy garden", "hot spring bath" and "forest bathing".

康复花园与城市绿地、公园存在着密切的联系，它不但具有美学和生态学范畴的功能，更具有减压及恢复身心健康的功能，对缓解城市快节奏生活带来的都市病症具有重要意义。身处一个好的康复花园，使用者可以通过感官与之交流，感受自然生命周期，增强自我认知，进而实现平和宁静的内心需求。著名的康复花园研究先驱、《人性场所——城市开放空间设计导则》作者克莱尔·库珀·马库斯曾说：康复花园不仅是一处地方，更是一个过程，这个过程即与自然互动，让自然给予身心放松的疗愈过程。如果能把康复花园的设计理念和方法融入城市绿地和公园，针对不同使用人群的身心康复需求进行精准设计，不仅能为使用者带来福祉，更能带来可观的经济效益和社会效益。同时，这也与构建人类卫生健康共同体的发展战略高度契合。

Closely related to parks and urban green space, healing garden not only has the aesthetic and ecological functions, but also has the functions of releasing pressure and helping one's physical and mental recovery. It is of great significance to alleviate the urban diseases brought about by the fast-paced urban life. In a good healing garden, users can communicate with it through their senses, feeling the natural life cycle, enhancing their self-cognition, and then acquiring their inner peace and serenity. Clare Cooper Marcus, a famous pioneer researcher of healing garden and author of *People Places: Design Guidelines for Urban Open Space*, once said that healing garden is not only a place, but also a process, the process of interacting with nature and the healing process of deriving physical and mental relaxation from nature. If the design principles and methods of healing garden can be incorporated into parks and urban green space, and if healing garden can be designed precisely to satisfy the physical and mental rehabilitation needs of different groups of users, healing garden can present the users not only with welfare, but also with considerable economic and social benefits. Meanwhile, it is highly consistent with our development strategy of constructing a Global Health community.

1.2　医养结合的康复花园
1.2　Healing garden combined with medical health care

　　康复花园或"大自然"何以对人体起康复作用？虽然目前鲜有公认的理论可以深入阐释其作用原理，但它的积极效应已经在历史研究及实践中被证实。目前，与医院结合的康复花园逐渐被认可，已经涌现出一批经典康复花园案例，如美国芝加哥植物园比勒康复花园、伊丽莎白与诺娜·埃文斯康复花园、梅西癌症中心康复花园、赛奇伍德失智患者康复花园、俄勒冈烧伤中心花园、约翰霍普金斯医院康复花园，中国北京积水潭医院康复花园等。

　　How can healing garden or "nature" contribute to the recovery of human health? Although there are rare recognized theories currently to deeply illustrate its working principles, its positive effects have been confirmed in the historical researches and practices. Healing garden combined with hospitals has gradually been recognized at present. There emerge a batch of classic healing garden cases, such as Buehler Enabling Garden of the Chicago Botanic Garden, Elizabeth & Nona Evans Restorative Garden, Massey Cancer Center Healing Garden, Sedgewood Dementia Commons Garden, Oregon Burn Center Garden, Healing Gardens of Johns Hopkins Medicine, and Healing Garden of Beijing Jishuitan Hospital, China.

1.2.1　康复花园作用机制
1.2.1　Mechanism of healing garden

　　康复花园理论奠基人罗杰·乌尔里希是最早展开医疗环境和疗效关系研究的学者，自 20 世纪 80 年代就专注于自然体验与人体康复之间的因果联系机制的研究，他通过一组病房窗景的对比试验证实了医疗环境中"大自然"的疗愈能力（Ulrich，1984）。他的主要成果压力缓解理论（SRT）和卡普兰夫妇的注意力恢复理论（ART）是目前公认的阐释康复花园作用机制的主要理论（Ulrich，1986；Kaplan，1995）。前者受"亲生命假说"的影响，认为个体处于压力状态会产生消极情绪，接触自然会阻断消极情绪，一旦脱离自然，人就会感到压力；后者认为人的大脑会因长时间处于主动注意模式而感到疲劳或紧张，当接触大自然时则会开启被动注意模式，以缓解主动注意模式造成的压力和紧张，最终获得恢复。关于康复花园作用机制的主要理论将在第 3 章详细阐述。

　　Roger Ulrich is the founder of healing garden theory and one of the earliest scholars to initiate the research into the relationship between the medical environment and the healing effect. Since 1980s, he has begun to focus on the research of the causal mechanism between nature experience and physical rehabilitation, and confirmed the healing capacity of nature in the medical context through the contrast method of experiment with ward window view

（Ulrich, 1984）. His major contributions, Stress Reduction Theory（SRT）, and Mr. and Mrs. Kaplan's Attention Restorative Theory（ART）, are currently the generally acknowledged major theories to interpret the working mechanism of healing garden（Ulrich, 1986; Kaplan, 1995）. The former is affected by "biophilia hypothesis", believing that an individual under pressure is sure to generate negative mood and that the contact with nature can prevent it. Once human beings are divorced from the nature, they will feel stressed. The latter insists that human brain will feel tired or nervous due to being in the active attention mode for a long time. In contact with the nature, human beings will switch on the passive attention mode to relieve the stress and tension caused by the active attention mode and to finally recover themselves. Chapter 3 will elaborate on the major theories about mechanism of healing garden.

1.2.2　康复花园的设计方法
1.2.2　Design methods of healing garden

目前，行业内较推行的康复花园设计方法是循证设计，将在第 3 章详细阐述。在循证设计中，医生、设计师、心理学家和使用者需要协同合作。设计方案需结合使用者的特点与需求，确定影响康复的环境因素，以科学试验和使用者康复指标数据为指导，进行精准设计。花园建成后，通过对使用者行为观察、问卷调查、恢复效果、住院时间、止痛药强度及剂量，甚至采用正电子发射断层扫描和功能性磁共振成像等医学方法，跟踪记录使用者大脑动态反应，以测评使用者的心理感受以及花园的康复功效，所获得的实证数据可用于指导其他康复花园设计。目前康复花园的循证证据来源主要包括：描述性研究、分析性研究、实验性研究和理论性研究。

Among healing garden design methods, the comparatively recommended is evidence-based design, in which doctors, designers, psychologists and users need to collaborate with each other. This design method will be discussed in detail in chapter 3. Based on the users' characteristics and needs, healing garden design need determine the environmental factors that may influence one's rehabilitation and be accurately targeted with scientific experiments and the users' rehabilitation indicator data as its guide. After the construction of healing garden, the users' psychological feelings and the healing effect of the garden can be evaluated through the user behavior observation, questionnaire, recovery effect, hospitalization time, painkillers' intensity and dose, and even through such clinic methods as tracing and recording the users' dynamic brain response via the positron emission tomography and functional magnetic resonance imaging. The obtained empirical data can be used to guide the design of other healing garden. At present, evidences of healing garden mainly originate from descriptive research, analytical research, experimental research, and theoretical research.

除了循证设计，也有专家提出亲生物设计、支持性花园设计等方法。亲生物设计以人的亲生物性为理论基础，通过设计契合并激发人的潜在生物本能，实现人向往自然、亲近自然并从自然中获得力量的本质；支持性花园设计以人的需求为出发点，体现人文关怀，通过缓解压力和不良情绪的设计助益康复。

In addition to evidence-based design, experts also propose other methods like biophilic design and supportive garden design. Based on the theory of human affinity for nature, biophilic design corresponds with and stimulates human potential biological instinct, trying to realize the human nature of yearning for nature, being close to nature and thus gaining strength from nature. Based on human needs, supportive garden design reflects humanistic care and contributes to rehabilitation through relieving stress and bad mood.

1.2.3　康复花园设计和营造的未来趋势
1.2.3　Future tendency in the design and construction of healing garden

现有康复花园所针对的使用者大多数还是医疗机构的患者。住院治疗期间患者的压力多来自服用药物、生活不便及对疗效的不确定等，康复花园的首要目标便是缓解患者压力。什么样的花园能最大化地实现这一目标？设计师、医生甚至患者自身都无法清晰表述，因此，循证设计数据和花园使用后评价数据的作用和意义尤其重要。目前，行业内急需积累实证数据、急需建立涵盖不同患者的康复花园质量评价指标和系统，使康复花园质量评价向专门化的方向发展，进而促使康复花园设计向着专业化、精准化、高效化方向发展。

The existing healing garden users are mostly patients in medical institutions. During hospitalization, patients are mostly under pressure from medication, inconvenient life and uncertain healing effect, and the primary goal of healing garden is to relieve their pressure. What kind of garden can maximize this goal? Designers, doctors and even patients themselves do not have a clear idea about it. Therefore evidence-based design data and users' occupancy evaluation data have a particularly important role and meaning. At present, it is urgent to accumulate empirical data, and establish quality evaluation index and system of healing garden targeted at different patients, so that its quality evaluation can develop in the direction of specialization and its design can then further develop in the direction of specialization, precision and efficiency.

不少专家指出，未来借助神经科学理论和技术有望进一步揭示康复花园的作用机制。

As many experts pointed out, the working mechanism of healing garden is expected to be further revealed with the help of neuroscience theory and technology.

1.3　可持续健康社区
1.3　Sustainable healthy communities

较之于医疗环境中的康复花园主要服务对象是医院人群，社区康复花园则是面向所有城市人群，其更能对整个城市环境的健康化作出积极贡献。通过构建可持续的、健康发展的社区，为广大城镇居民提供绿色康养活动空间，有望引领都市人群的健康生活方式，缓解因环境不良诱发的都市病症等。

Healing garden in the medical environment is mainly targeted at patients, while community healing garden is open to all the urban population so that it can better contribute to the healthy urban environment. The construction of sustainable and healthy community can present the urban citizens with green rehabilitation activity space, leading them into a healthy lifestyle and alleviating urban diseases caused by the poor environment.

1.3.1　可持续健康社区的概念
1.3.1　The concept of sustainable healthy communities

相较于那些不健康社区而言，人们可以在可持续健康社区体会健康和幸福的感觉，降低对医疗服务的需求。在 2013 年的世界设计与健康大会上，许多研究报告详细阐述了景观的积极作用（Gayle，2014）。从金融学到心理学等领域的专家们，谈到了他们在采取预防性保健措施方面的成功经验。已有学者认为不良的城市环境长期以来一直是导致人们不健康的根源，许多研究已经明确了自然环境和健康提升之间的积极关系，例如伊格纳·蒂那娃等人表明"从生态角度考量，通过种植乡土植物来绿化城市，也是生态系统和公民整体福祉的重要组成部分"（Ignatieva，2008）。

In contrast with unhealthy communities, a sustainable healthy community can help people experience a sense of well-being and reduce their demand for healthcare services. At the 2013 World Congress on Design and Health, a number of researches elaborated on the benefits of landscape (Gayle, 2014). Experts in the fields ranging from finance to psychology mentioned their successful experiences in adopting preventive healthcare implementations. Not only did some scholars argue that poor urban environments have long been the source of ill health, but also many studies have identified the positive relationship between the natural environment and improved health, for example, Ignatieva et al. stated that "from an ecological point of view, urban greening by planting native species is also an important integrated part of ecosystems and the overall human well-being" (Ignatieva, 2008).

要想营造一个可持续的社区，需要实现公民和生态系统之间的平衡，不能局限于现有资源，因为这将影响未来几代人的生存能力。美国布伦特兰委员会将可持续发展定义为："可持续发展是指既满足当代人的需要又不损害后代人满足其需求的能力。"所以，当我们开始建造康复花园时，是在做一件兼顾人类和环境健康的益事。

A sustainable community needs us to create the balance between mankind and ecosystems instead of just being limited to the resources available today, due to its potential influence on the future generations' survival ability. The Brundtland Commission defined sustainable development as "development that can meet the needs of contemporary people without compromising the ability of future generations to meet their own needs". When we start building healing gardens, we are contributing to human health and healthy environment.

1.3.2 可持续健康社区的构成要素
1.3.2 The constitution of sustainable healthy community

健康社区的发展核心就是可持续性，包括环境的可持续性与经济的可持续性。要确保经济的可持续发展，健康的人群是第一要素，而环境的可持续决定了人群的健康水平。因此，要想实现真正的可持续发展就要同时兼顾以下两个元素：①符合成本效益的设计；②以自然为主导的社区环境提升。尊重、保护原有地貌景观和本土植物可以大大降低建造成本，利用植物的生物多样性来体现社区的自然风貌也更能增强景观的观赏性和实用性。

The core of developing a healthy community is its sustainability, including environmental and economic sustainability. To ensure economic sustainability, the first element is the healthy population while environmental sustainability can determine the level of health of the population. In consequence, the realization of true sustainable development needs the following two simultaneous elements: ① the cost-effective design; ② nature-oriented community environment improvement. The construction cost can be greatly reduced through the respect and the conservation of the original landscape and native plant species. In addition, the landscape's ornamental and practical values can also be enhanced by applying plant biodiversity to exhibit the community's natural landscape.

思考题

1. 康复花园与普通城市绿地的区别是什么？
2. 康复花园的特征有哪些？
3. 为什么要构建可持续健康社区？

Questions

1. What is the difference between healing garden and ordinary urban green space?
2. What are the characteristics of healing garden?
3. Why should sustainable and healthy communities be built?

第 2 章

康复花园设计哲学观

Chapter 2 Design Philosophy of Healing Gardens

　　设计哲学是 21 世纪新兴的学科。设计的历史可以追溯到石器时代，而哲学总是与知识、智慧、思考、逻辑、分析相关联，以思考人生、世界、时空和宇宙等为终极目标。如果我们将造园活动、园林设计看作一个独立世界，设计师为其分析需求、建立模型、捕捉内在运转的规律、设计合理的架构，不就是在做哲学家每天在做的事情吗？规律和本质，在设计的世界中被我们创造，反过来又指导着我们，规定着设计的方向，这就是设计哲学。关于康复花园，虽然目前还未建立起系统的设计哲学，但康复花园的设计终究是人居环境的设计，故而从顶层逻辑而言离不开"人居背景、人居活动和人居建设"三方面辩证统一的哲学观（刘滨谊，2016）。因此，在康复花园的设计过程中，强调设计哲学观亦是本着人类对环境认知以及对人居环境构建的进步去拓展的。

　　Design philosophy is an emerging subject in the 21st century. The history of design can be traced back to the Stone Age, and with the ultimate goal of thinking about life, world, time and space and the universe, philosophy is always associated with knowledge, wisdom, thinking, logic and analysis. If we regard garden-design activities or landscape architecture as an independent world, designers analyze the design requirement to establish models, and then capture the internal patterns of operation to design reasonable structures, which is exactly what philosophers do every day. Patterns and essence are created by us in the design world and, in turn, guide us and determines the direction of design, which is design philosophy. For healing gardens, although no systematic design philosophy has been established, the design of the healing gardens is ultimately the design of human settlements, so from the top-level logic, it cannot be separated from the dialectical and unified philosophical view of "human settlement background, human settlement activities and human settlement construction" (Liu Binyi, 2016). In the process of designing the healing gardens, we emphasize that design philosophy is also based on the progress of human cognition of the environment and the construction of human settlements to expand.

科学技术的迅猛发展，在创造世界经济奇迹的同时，也使地球的资源和环境遭到前所未有的破坏。地球、人类固有的平衡被打破，生态安全已在全球部分地区亮起红灯，代表人类与环境关系的生态健康、人类自身的健康正在受到严重威胁。堪称人与植物良性互作关系典范的"园艺康健和康复花园"，是生态文明建设在人类健康照护方面的体现。同时，康复花园是风景园林专业中近年来新兴的融风景园林、园艺学、医学、心理学、社会科学等知识于一体的方向，是新工科、新农科、新医科和新文科交叉的领域，围绕着风景园林与健康生活的社会需求，康复花园的设计建造可谓将生态文明与健康照护落到了实处。因此，本章立足风景园林三元论基本思想，主要从康复花园设计与人类健康关系中需要重点思考的包容性、平衡性展开对康复花园设计哲学观的探讨。

With the rapid development of science and technology, though the world economic miracle was created, the unprecedented destruction of the earth's resources and environment was caused at the same time. The inherent balance between the earth and human beings was broken, and the indicators of ecological security have been flashing red in some parts of the globe. Ecological health which represents the relationship between human beings and the environment as well as the health of human beings is under severe threat. "Horticultural therapy and healing garden" can be regarded as a model of the benign interaction between human and plants, which embodies the eco-civilization construction in human nursing care. At the same time, healing garden is an emerging direction of the major of landscape architecture in recent years, integrating the knowledge of landscape horticulture, medicine, psychology and social sciences. It is an interdisciplinary field of new engineering, new agriculture, new medicine and new arts. Around the social needs of healthy life and landscape architecture, the design and construction of healing gardens can be described as the implementation of ecological civilization and human nursing care. Therefore, based on the basic idea of Ternary Theory of landscape architecture, this chapter mainly discusses the philosophical view of healing garden design from the inclusiveness and balance that need to be emphasized in the relationship between healing garden design and human health.

2.1　风景园林三元论理论思想
2.1　Thought of Trialism Theory of landscape architecture

三元论理论思想由我国学者刘滨谊教授提出，其背景是立足风景园林是人居环境的重要组成部分之一，其设计建造规律亦遵循人居环境三元论的理论基础。人居环境三元论是指将人居环境理论研究分为人居背景、人居活动和人居建设三大要素，并在这三大要素辩证统一的叠加研究之下，指导人居环境建设达到从解决问题到实现理想环境的目标(刘滨谊，2016)(图2-1)。该理论认为，当代人居环境以自然界环境、农林环境和生活环境三者为存在基础，其中包含自然环境、农林环境、生活环境三类空间环境，以及各类环境中所具有的各类资源和生态循环等，它们维持着人类的基本生存，是人居环境存在的必要前提，此部分研究为人居背景；人类利用人居环境进行的各类居住、聚集和游历活动是人居环境的主体及其表现形式，此部分研究为人居活动；集中体现人类在各类空间中建设活动成果的建筑、城乡、风景园林与景观等，则是人居环境的客体及其表现形式。

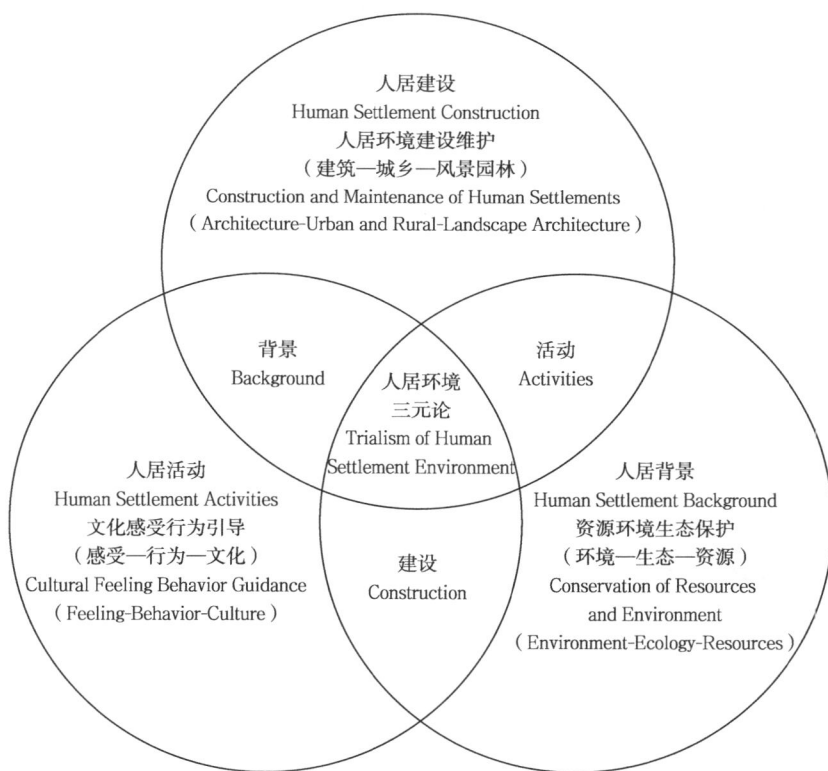

图 2-1 人居环境三元论示意图(刘滨谊，2016)

Fig. 2-1 Schematic diagram of Trialism of Human Settlement，Inhabitation and Travel Environment Studies(Liu Binyi, 2016)

Trialism Theory was proposed by Professor Liu Binyi, a Chinese scholar. Its background is based on the fact that landscape architecture is an important component of human settlement environment, and its design and construction laws also follow the theoretical foundation of the trialism of human settlement, inhabitation and travel environment studies. The trialism of human settlement environment refers to the division of theoretical research on human settlement environment into three major elements: human settlement background, human settlement activities, and human settlement construction. Under the dialectical and unified superposition study of these three elements, it guides the construction of human settlement environment to achieve the goal of solving problems and achieving an ideal environment (Liu Binyi, 2016) (Fig. 2-1). This theory believes that contemporary human settlements are based on the natural environment, agricultural and forestry environment, and living environment, which include three types of spatial environments: natural environment, agricultural and forestry environment, and living environment, as well as various resources and ecological cycles in each type of environment. They maintain the basic survival of human beings and are necessary prerequisites for the existence of human settlements. This part of the research is the background of human settlements. The various residential, gathering, and touring activities carried out by humans using the living environment are the main body and manifestation of the living environment, and this part of the research is related to human settlement activities. Buildings, urban and rural areas, scenic gardens, and landscapes that reflect the achievements of human construction activities in various spaces are the objects and manifestations of the living environment.

可见，在该理论体系中，人居活动是实现理想环境构筑过程中不可或缺的一环，其为康复花园与园艺康健的密不可分提供了强有力的理论支撑。"园艺康健"的实施细则就是人类在实现康复花园疗愈目标过程中的一种人居活动。

It can be seen that in this theoretical system, human settlement activities are indispensable in the process of constructing an ideal environment, which also provides strong theoretical support for the inseparable relationship between rehabilitation gardens and horticultural therapy. The implementation of horticultural healing is actually a human settlement activity in the process of achieving the goal of rehabilitation.

2.2　设计的包容性
2.2　Inclusiveness of design

包容性是康复花园的一个重要特征，是指要求花园能够使不同年龄阶段、不同种族、不同身份、不同能力的人聚集在同一个环境中且感到舒适。重要的是身处这个环境，他们能感觉自己与其他人一样，受到相同程度的重视。例如，大部分的花园是无法为肢障人群提供园艺操作空间的，但是，当肢障人群在康复花园中，他们就能够享受到与普通人一样的园艺劳作带来的快乐和尊严。

Inclusiveness is an essential feature of healing gardens, which refers to the requirement that people of different ages, races, identities and abilities can gather in the same environment and feel comfortable. The important thing is that in this environment, they can feel that they are valued to the same degree as others. For example, most of the gardens can not provide gardening space for the disabled, but when the disabled are in the healing gardens, they can enjoy the same pleasure and dignity brought by gardening as ordinary people do.

2.2.1　包容性设计思维的培养
2.2.1　Cultivation of inclusive design thinking

包容性设计要求充分认识不同个体之间存在的差异，针对不同人群的特征，立足于"以人为本"的理念，通过以软质景观为主导的空间营造去满足这种差异性。我们需要一种能够满足某一群体特殊需求的设计体系，同时还要具有普遍吸引力，这就是一种包含整体性思维的成果导向方法，要求我们能够摒弃世俗成见，积极地去识别、鼓励并接受个体差异，同时还要考虑生活固有的适应性，这样就能够培养包容性思维，便可在人类所能达成的社会关系中进行景观设计，这种社会关系决定人类在各自所处环境中的行为。例如，近年来研究发现，与生活方式相关的疾病和安全等问题，可以通过提升环境健康得以解决。越来越多的科学证据表明，在工业化的城市，家庭和其他建筑物内的空气污染比室外更严重。为了应对这种情况，公共卫生从业者、建筑师和城市规划师与景观专家必须展开深度合作，共同设计有吸引力的环境，为人们提供康复花园，鼓励人们花更多的时间进行户外活动；无论是在办公场所、学校、护理机构，抑或在高层公寓楼或独立的家庭住宅房，绿色空间都可以为心灵和身体提供精神食粮。所以，康复花园的设计是极具包容性的循证设计，这种设计的方法打破了"甲方乙方"单纯的双边关系，要求从设计之初就涵盖"投资方""研究方""设计方"和"使用方"的多边关系。只有从多角度的视点和需求切入，才能够设计出真正具有包容性的康复花园，以此去引领人们健康的生活习惯和行为方式。

Inclusive design requires us to fully understand the differences between different individuals, and then, according to the characteristics of different groups, to meet the differences by creating a soft-landscape-dominated space on the basis of the "people-oriented" concept. We need a design system that can meet the special needs of a certain group and that has universal appeal, which is an outcome-oriented method with a systematic thinking pattern. It requires us to abandon stereotypes and actively identify, encourage and accept the individual differences, and also take the inherent adaptability of life into consideration, so that we can develop inclusive thinking to practice the landscape architecture in the social relationship that people have forged, which determines people's behavior in their own environments. In recent years, it has been proven that lifestyle-related diseases and safety issues can be addressed through research in environmental health. More and more scientific evidences show that, in industrialized cities, the air pollution at home or in other buildings is more serious than that of the outside. In order to cope with this situation, public health practitioners, architects, urban planners and landscape experts must have in-depth cooperation. They need to design attractive environments together to provide healing gardens for people and encourage people to spend more time outdoors. Green spaces can always give your body and soul the food for thought, no matter whether it is in workplaces, schools, nursing care facilities, the high-rise apartment buildings or even in detached family houses. Therefore, the design of healing gardens is the most inclusive evidence-based design, which breaks the simple bilateral relationship between "Party A and Party B", and which requires that the multilateral relationship including "investor", "researcher", "designer" and "user" should be covered at the beginning of the design. Only from multiple perspectives and needs can we design a truly inclusive healing garden to lead people's healthy habits and behaviors.

2.2.2 包容性设计的要点
2.2.2 Key points of inclusive design

（1）深入调研

（1）In-depth research

当研究一个新的设计方案时，为确保设计具有包容性，接触使用者并展开深入调研是非常重要的；可以采取问卷、面谈等方式。调研问卷的设计必须根据设计主题来展开，如提问：我想知道你最喜欢让这个空间变成什么样子？你会改变什么？你理想中它的样子和感觉是什么样的……调查者可以耐心地直接问，也可以通过互联网、信件等方式发放问卷，这过程可能漫长而艰难，但通常是获取最为周到方案的最佳途径。

When looking into a new design, it is very important to contact users and conduct some in-depth research to ensure that the design is inclusive. The research method include questionnaire, interview and so on. Furthermore, the design of the questionnaire must be based on the theme you designed. For questions like "I wonder what you prefer the space to be like." "what will you change?" and "what do you think its ideal appearance looks like and feels like?" You can ask directly with patience, or you can send questionnaires through the Internet, letters and other ways. In addition, you need to know that you should make preparations because it may be a long and slow process, but it's usually the best way to achieve the most comprehensive design.

当与残障人士接触时，有必要在第一次客户会议之前研究使用者的健康状况。如在设计特殊教育环境住宅时，共处一室的家庭成员各自的特殊需要都是待解决的问题。设计者需要尊重并知道他们的意见和需求，这将为包容性设计解决方案提供信息。附录 1 提供了与残疾相关的术语表，有利于了解使用者的需求；如果客户是大型医院、学校或市政当局的管理团队成员，还有一个重要的环节是要与空间的实际使用者进行深入交谈，以获取这些实际使用人群的第一手资料。

When contacting the disabled, it is necessary to study the users' health before the first client meeting. For instance, when designing residential environments for special education, the specific needs of family members in the same room are all issues to be addressed. We need to respect and know their opinions and needs, which will offer us inclusive design solutions. Appendix 1 provides a glossary of terms related to disability, which helps us understand the needs of users. If the client is a member of the management teams of large hospitals, schools or municipal authorities, it is also important to have an in-depth conversation with people who actually use the space, so as to obtain the first-hand information.

（2）以人为本

（2）People-oriented concept

在设计有助于促进健康与愉悦的景观时，首先需要从经验创建角度对设计项目进行评估，评估内容上至总体规划阶段、下至详细的小规模设计。以人类健康为本源的设计重心在于创建能刺激人们产生亲生物感觉的空间。建立人与自然的联系，无论公用还是私用的空间都需要令人感到舒适、较强的可达性，并可接纳不同年龄及能力的个人。让患有失智、中风或脑损伤的老年人，患有多动症、哮喘、脑性麻痹、唐氏综合征、癫痫、肥胖症、脊柱裂、听力和视力障碍的儿童，以及其他先天或后天残疾的人和他们的家庭，都可以从康复花园中获得益处。

When designing a landscape that helps to promote health and happiness, we need to first assess the design project from the perspective of experience creation, from the master plan down to the detailed small-scale design. The focus of design based on human health is to create a spaces that can stimulate us to have a biophilic feeling. To establish a connection between man and nature, both the public and private spaces should be comfortable, and accessible, and can accommodate all individuals of different ages and abilities. People including the elderly suffering from dementia, stroke or brain injury, children with attention deficit hyperactivity disorder (ADHD), asthma, cerebral palsy, Down syndrome, epilepsy, obesity, spina bifida, hearing and visual impairment and others with congenital or acquired disabilities and their families can all benefit from the healing gardens.

2.3　设计的平衡性
2.3　Balance of design

2.3.1　平衡的重要性与失衡之痛
2.3.1　The importance of balance and the lesson of imbalance

从人类影响和改变自然生态过程的角度来看，环境问题始终伴随着人类的发展。16~17 世纪以来，特别是第二次世界大战以后，由于工业化、人口膨胀和城市化进程的快速发展，人类社会的经济活动对地球环境的干扰愈演愈烈，从而导致了严重的环境污染和生态破坏。20 世纪以来，人类在创造世界经济奇迹的同时，正以惊人的速度破坏地球的生态平衡。

From the perspective of human impact on and change to natural ecological process, environmental problems have always been accompanied by human development. Since the 16th and 17th centuryies, especially after the Second World War, due to the rapid development of industrialization, population expansion and urbanization, the interference of human society's economic activities in the earth's environment has become more and more intense, resulting in serious environmental pollution and ecological damage. Since the 20th century, though human beings have created the world economic miracle, they are destroying the ecological balance of the earth formed in billions of years at an amazing speed.

失衡的人地关系让社会的健康陷入低谷。城市的发展让房屋替代了森林，大地不再透气，人们与土地的关系永远隔着一层坚硬的混凝土；街道、大院的消失，让城市完成了陌生化的进程，人们沦落为孤独的居住者，不断隆起的钢筋水泥建筑，让城市简化为超市、网吧、咖啡屋、小区和办公楼；人们的周围是一个个目的性明确、无闲情逸致的居住空间。人际关系也因此变得冷漠。

For the imbalanced relationship between people and land, the health of society has fallen into a slump. Due to the city development, the house replaces the forest, the earth is no longer ventilated, and mankind is always kept isolated from the land with a layer of rigid concrete. The disappearance of streets and yards made the city complete its process of de-familiarization. As a result, people have been reduced to lonely residents. Because of the rising steel-and-concrete "monster", the cities are simplified into supermarkets, Internet cafes, coffee houses, residential areas and office buildings. People are surrounded by living spaces with a clear purpose but without any leisure. People live in a lonely and noisy way willfully.

失衡的人地关系与天人合一思想背道而驰，2020 年一场突如其来的新冠肺炎疫情让我们每个人都深刻领悟了失衡的痛苦。每个人都在盼望疫情早日过去，盼望我们的生活早日回归以往的平静、衡稳。例如，在平衡大师志田美代子表演的羽毛平衡术中，就能让观众体会到平衡的美感，体会维系平衡需要的认真谨慎和小心翼翼；体会一个平衡的系统中，每一个组成都是至关重要的，哪怕轻如鸿毛，也是维系平衡的重要因素！如此这般，在康复花园的设计体系中，平衡也均由每个细节造就。

The imbalanced relationship between people and land has run counter to the idea of harmony between man and nature. The year 2020 witnessed that a sudden COVID-19 forced each of us to learn the lesson of the imbalance. All everyone wants is for the epidemic to be over and for our life to return to calm, balance and stability as it was. For instance, in the Sanddorn Balance performed by Balance Master Miyoko Shida Rigolo, the audience can realize the beauty of balance, realize how careful and cautious the master is to maintain the balance, and realize each part is very important in a balanced system, and even if it is as light as a feather, it plays an indispensable role in maintaining the balance. Because of this, in the design system of healing gardens, its balance is created by every detail.

2.3.2　康复花园的平衡要素
2.3.2　Balance elements of healing gardens

本着呵护健康的初心，康复花园的设计倡导重建平衡的人地关系与人际关系，因为生态的健康需要平衡、环境的健康需要平衡、人体的健康更需要平衡。当今世界，环境污染、精神负荷超重和活动不足等严重威胁着公共健康。人们对居住空间的要求从绿化到美化，再到当今的园林环境健康化，所以，康复花园的设计建造就是在重塑城市园林环境与健康关系的过程中，去不断追求平衡。可从以下四个方面来探讨平衡的要素。

With the original aspiration of caring for health, the design of healing gardens advocates the reestablishment of a balanced human-and-land relationship and human relationship, because ecological health needs the balance, environmental health needs the balance, and human health needs the balance more. In the world today, these issues like environmental pollution, excessive mental stress and physical inactivity pose serious threats to public health. People's requirements for living space are from greening to beautification, and then to today's healthy landscape environment. Therefore, the design and construction of healing gardens is to constantly strive to achieve the balance in the process of reshaping the relationship between urban landscape environment and health. We will discuss the balance elements of healing gardens from the following four aspects.

（1）建筑与绿地的平衡——容积率、绿地率

（1）Balance between building and green space—floor area ratio and greening ratio

容积率，是指一个小区的总建筑面积与用地面积的比率。对于开发商来说，容积率决定地价成本在房价中占的比例，而对于住户，容积率直接涉及居住的舒适度。如果说建筑是凝固的音乐，那么植物就是流动的旋律；如一个良好的居住小区，高层住宅容积率应不超过5，多层住宅应不超过2，广场绿地率应高于35%。城市规划法规体系下，编制的各类居住用地控制性详细规划中，所列的容积率见表2-1［其他更多参数请参考资料《城市绿地分类标准》（CJJ/T 85—2017）］。

表 2-1 不同居住用地容积率要求

Tab. 2-1 Requirements for the FAR of different residential land

居住用地类型 Types of residential land	容积率 FAR	居住用地类型 Types of residential land	容积率 FAR
独立别墅 detached villa	0.2~0.5	11 层小高层住宅 11-storey high-rise residence	1.5~2.0
联排别墅 townhouse	0.4~0.7	18 层高层住宅 18-storey high-rise residence	1.8~2.5
6 层以下多层住宅 less than 6-storey residence	0.8~1.2	19 层以上住宅 residence with more than 19 storeys	2.4~4.5

Floor area ratio(FAR) refers to the ratio of a building's gross floor area to the size of the piece of land upon which it is built. For developers, FAR determines the proportion of land price cost in housing, while for residents, FAR is directly related to the comfort level of their living environment. If we see the architecture as the solidified music, we can say that plants are flowing melodies, for example, a good residential community should be like this: the

FAR of its high-rise residence should not exceed 5, multi-storey residence should not exceed 2, and greening ratio of the square should be higher than 35%. Under the urban planning regulatory framework, the FAR listed in various types of residential land control detailed plans are shown in Table 2-1 [For additional parameters, please refer to *Classification Standard for Urban Green Space* (CJJ/T 85—2017)].

（2）造园手法的平衡——山石、水体、植物、建筑
（2）Balance of gardening techniques: rocks, water, plants and buildings

园林体现了人类改善自身居住环境的意识觉醒，也是标志人类文明美学进步的人工自然。一座优美的园林是一个协调的环境空间。就造园形式而言，它是由山石水体、树木花草、楼台亭阁等要素组合成的一个综合艺术体系。利用不同造园要素的特性，采用平衡的造园手法，方能和谐组合、顺乎自然、改进自然，建造出或使人积极向上或使人轻松舒畅的艺术境地。所以，造园的实质是创造有利于生态平衡的、修身养性的环境。一方面，它必须遵循自然界内在的发展规律，即合律性；另一方面，又要使人在生产实践过程中满足其改善自身生活的需要，即合目的性。这种内容与形式的统一，目的性与规律性的统一，就是造园的基本原则。而各造园要素的和谐平衡是落实这个基本原则的第一步。

Landscape is not only the awakening of human consciousness to improve their living environment, but also the artificial nature which marks the progress of aesthetics of the human civilization. A beautiful landscape must be a harmonious environmental space. In terms of landscape design form, it is a comprehensive art system composed of rocks, water, trees, flowers, grass, pavilions and other key elements. Only by making use of the characteristics of different elements and adopting balanced techniques of landscape design, can we create an inspiring and relaxing world of art with a harmonious combination as well as a nature-based method. Therefore, the essence of landscape design is to create an environment conducive to ecological balance and rest-and-recuperation. On the one hand, it must follow the inherent law of nature, that is, the conformity to regularity; on the other hand, we should help people meet the needs of improving their life in the process of production practice, that is, the conformity to purpose. The unity of content and form and the unity of purpose and regularity are the basic principles of landscape design. The first step to implement these basic principles is to balance the elements of landscape design harmoniously.

如何做到造园手法的平衡呢？这需要设计者对造园要素的合理认知与利用。首先各造园要素自身要和谐统一，主要表现在要素的形态与质地的统一，根据材料自身的特质合理利用；其次要考虑要素之间关系的和谐统一，即各要素必定是相称的、有秩序的。

How can we achieve the balance of landscape design techniques? It requires designers to have a reasonable understanding and utilization of elements of landscape design. First of all, the elements of landscape design should be harmonious and unified, which is mainly reflected in the unity of elements' form and texture and the rational use of the materials according to their own characteristics. Secondly, we should consider the harmonious unity of the relationship between the elements, that is, the elements must be proportionate and orderly.

总之，花园中的山石、水体、植物、建筑等景物须有一定的体量和比例尺度。各景物不能过分突出和独立，须围绕造园的主题，以中心景物为主体，其他部分按照一定的比例分布。

In a word, the rocks, water, plants and structures in the garden must have a certain volume and scale. Instead of being too prominent and too independent, each part of the scenery should be centered on the theme of landscape design, with the central scenery as its main body and the other parts in a certain proportion.

（3）植物种类的平衡——生物多样性
（3）Balance of plant species——biodiversity

康复花园和其他类型的园林相比，造园的主要元素为植物，因此对植物种类的要求就相对高。就园林植物多样性的要求而言，全球已发现和描述的维管植物共 37.6 万余种（钱宏，2022），据英国皇家园艺学会（RHS）网站公布的植物名录检索：其根据观赏特性进行分类的植物涵盖 7.2 万余种，可用于城市绿地及庭院的植物 3.8 万余种；我国是世界上物种多样性最丰富的国家之一，据中国科学院昆明植物研究所官网发布，我国拥有 3.9 万余种高等植物，其中原产的观赏植物种类超过 7900 种；但城市绿化常用观赏植物仅有 2000 余种（李先源，2018）；这和世界"园林之母"的美称大相径庭。失衡的园林植物配置，让各城各地的绿地景观千篇一律、甚无新意，这也是近些年业界呼吁亟待改善的地方。

Compared with other types of gardens, plants can be regarded as the main element of healing gardens, so the requirements for plant species are relatively high. As far as the diversity of garden plant species is concerned, according to statistics, there are approximately 376 000 species of vascular plants have been discovered and described globally (Qian Hong, 2022). According to the plant database published by the Royal Horticultural Society (RHS) of the United Kingdom, around 72 000 species are classified based on ornamental characteristics, with about 38 000 species suitable for urban green spaces and gardens. China is one of the countries with the richest biodiversity in the world. As reported by the official website of the Kunming Institute of Botany, Chinese Academy of Sciences, China is home to over 39 000 species of higher plants, including more than 7900 species of native ornamental plants. However, only about 2000 species of ornamental plants are commonly used in urban landscaping (Li Xianyuan, 2018), which is quite different from the reputation of "the Mother of Gardens". Due to the imbalanced allocation of landscape plants, landscape design across China's cities is stereotyped and lacks originality, in which the industry has been calling for prompt improvement in recent years.

除了遵循因地制宜、适地适树的基本原则进行康复花园的植物配置，还需要遵循因人而异、对症配置的原则进行康复花园植物种植设计。不同的人群、不同的康复目的，需要通过不同的植物种类和配置手法来完成种植设计。例如，最常见的就是针对盲人使用的花园，需要采用大量芳香植物和可以吸引鸟类的植物，让盲人通过嗅觉和听觉感受花园的疗愈功能；而针对肢障人群，可以通过丰富的色彩和造型营造视觉饱和度高的花园；针对儿童好奇好动的特点，要选择充满趣味性和较高安全性的植物。传统的中医理论和最新的科学证据表明，不同的情绪问题亦可以通过不同的植物获得对症改善，这也是亟待深入研究的领域。

In addition to following the basic principles of adjusting measures to local conditions and planting suitable trees for the local area, we need to follow the principles of individual-specific and disease-specific plant allocation in garden plant design. Treating different people with various rehabilitation purposes, we need to use different plant species and allocation tactics to complete the planting design. For example, the most common garden is for the blind. We need to use a large number of aromatic plants and plants that can attract birds, so that the blind can feel the healing power of the garden through their sense of smell and hearing; for the disabled people, we can create a garden with high color saturation by rich colors and sculptures; according to children's characteristics of being curious and active, we should choose amusing and safer plants. Traditional Chinese medicine theory and the latest scientific evidences show that different emotional problems can be improved accordingly by different kinds of plants, which is also a new field requiring further research.

（4）设施功用的平衡——人性化设计
（4）Balance of facility function——human-friendly design

人们在游园的时候经常会有一些体验，即诟病不少设施中看不中用。究其原因，主要是因为设计之初可能只注重该设施的外形是否美观，而忽略了其功能的表达；再者是因为体现功用的设计往往有一定难度，需要设计师的精心思考与设计。最常见的一个例子就是"汀步"，本来汀步的出现源于在水面设置道路，但是近现代的园林使用了大量的"旱汀"，即在地面铺设汀步。铺设在园林地面弯弯曲曲、星星点点的汀步确实好看，增加了庭院的意趣，但是游人却很难找到一个合适步长的汀步，不是太长就是太短。因为每个人的步长是不一样的，汀步必然不可能满足所有人的舒适感，所以早些年有人就提出采用计算人体的平均步长的方法来设计汀步的距离，然而"平均"的理论在汀步的设计上既不平也不均，这样的结果更是让汀步难以下脚，成了名副其实的摆设。那么，从人性化舒适度的角度考虑，是否"汀步"这种路面设施就没有办法做合理设计？答案显然是否定的，我们不仅可以继续在康复花园中设计汀步，而且可以把汀步设计得比较合理。只是需要换个角度思考问题，换个方法去做设计。汀步虽然不能适应每个人的步长，但是汀步却可以让每个人在路过的时候有等距的步长，换言之，汀步可以起到引导游人步行速度的作用。例如，在一个需要放慢脚步才可以仔细欣赏的花坛前设置短距汀步，即无形中告诉游人这里的景观要细看；而在一个需要快速通过的景观节点，不妨设计长距汀步，让人大步前进，甚至设计让人跑起来的汀步。这样汀步的功能就落到了实处。

It is unavoidable that we can make complaints about many facilities when we visit a garden. Firstly, the main reason is that only the appearance of the facility is valued at the beginning of the design, but the function is ignored. Secondly, function-focused design is often more difficult and requires careful thinking and design. One of the most popular examples is step stones (stepping stones on water surface, also named "Tingbu" in Chinese). At the beginning, Tingbu originated from setting roads on the water surface, but modern gardens used a lot of "Hanting(step stones pathway on the ground)", that is, laying stones on the ground. It is a similar and shared experience that the winding and dotted stone pathway on the ground is really beautiful, and increases the charm of the courtyard. However, it is difficult for visitors to get a suitable step size. It is either too long or too short. Because everyone's step length is different, it is impossible to satisfy everyone. In the early years, it is proposed to design the step length of "Tingbu" according to the average step length of human beings. However, the "average" theory is inapplicable in the design of step length of "Tingbu". As a result, Tingbu becomes a real decoration for it is more difficult to walk on. From the perspective of human comfort, isn't there any reasonable way to design the "Tingbu"? The answer is "absolutely yes, there is". We can continue to design Tingbu in the healing gardens in a more reasonable way. We just have to think differently about it, and find a new approach to design it. Although Tingbu isn't suitable for everyone's step length, it can ensure that everyone can walk by with equal step length. In other words, Tingbu can guide the walking pace of tourists. For instance, setting a short-distance Tingbu in front of a flower-bed needs the visitors to slow down to fully appreciate it, which implies that visitors should look closely at the landscape here. In a landscape node that needs to pass quickly, we had better design a long-distance Tingbu to help people stride forward, or even to design a Tingbu to make people run. In this way, the function of Tingbu has been implemented.

以上汀步设计案例是较为典型的道路设施设计。总之，在进行康复花园设计时，每一处设施都要根据功能的需要进行设计。首先需要思考花园解决什么问题，什么样的设施可以解决这个问题；接下来再思考这个解决问题的设施该如何设计才合理。这样才能让康复花园设施真正起到作用，而非中看不中用的摆设，真正让设施与其功用达到平衡。

The above examples are typical road facility designs. In conclusion, when we design the healing gardens, every facility should be designed according to the functional needs. First of all, we need to address what problems we need the garden to solve and what kind of facilities can solve this problem. Next, we need to think about how to design the facilities reasonably to solve it. Only in this way, can beautiful and practical facilities of the healing gardens, really work to reach a true balance between these facilities and their functions.

思考题

1. 康复花园设计为何要强调设计哲学观？
2. 康复花园的包容性设计体现在哪些方面？
3. 如何做到康复花园设计的平衡性？

Questions

1. Why does the design of healing gardens emphasize design philosophy?
2. What are the aspects of inclusive design of healing gardens?
3. How can the balance in the healing garden design be achieved?

第3章

康复花园循证设计的理论基础

Chapter 3　The Theoretical Basis of the Evidence-Based Design of Healing Gardens

　　康复花园的概念提出虽然只有短短数十年，加之其跨学科研究的特殊性，研究理论深度相对薄弱，但是就人类文明的发展史而言，康复花园的实践运用却拥有上千年的历史，在整个发展过程中也积累了前人关于人居环境的经验和智慧。本章从园林环境与健康的关系剖析入手，围绕园林植物对人体身心健康的作用展开探讨，简要梳理了康复花园在循证设计过程当中，从实践需求到学术研究所遵循的基础理论知识。

　　The concept of the healing gardens have been put forward for just a few decades, coupled with the particularity of its interdisciplinary research, so the research theory is relatively weak in its depth. But healing gardens have been carried out and applied for thousands of years from the perspective of the development history of human civilization, and throughout their development history, the experience and wisdom of predecessors on human being's settlement environment have been accumulated. This chapter starts with the analysis of the relationship between the garden environment and health, and probes into the effects of garden plants on the physical and mental health. It briefly sorts out the basic theoretical knowledge, with which practical demands and academic researches comply, of healing gardens in the process of the evidence-based design.

3.1　园林环境与健康的关系
3.1　The relationship between the garden environment and health

　　环境的优劣与人的身心健康是息息相关的。实践证明，人们通过接触自然环境、观赏植物可以达到放松心情、减轻压力，进而保持身心健康的目的。古今中外，医学、心理学、社会学、植物学、风景园林学、园艺学等不同领域的学者从来不乏对植物环境与人类身心健康关系的研究，他们从不同的角度出发，利用不同的方法和手段，共同推进这一领域的发展。

The pros and cons of an environment are closely related to the physical and mental health. Practice has proven that people can lighten up and reduce stress by engaging with the natural environment and enjoying the sight of plants, thereby maintaining their physical and mental health. Scholars at all times and in all over the world in different fields such as medicine, psychology, sociology, botany, landscape architecture, and horticulture have never lacked researches on the relationship between the plant environment and physical and mental health. They start off from different angles and use different methods and means to jointly promote the development of this field.

3.1.1　健康的概念
3.1.1　The concept of health

　　世界卫生组织对健康的定义指出：健康不仅是没有疾病，也是躯体、心理与社会适应能力的完好状态。郑丽在《康健园艺与康复花园》一书中对健康的外延做了拓展性探讨，认为健康可以包括"个人健康、社会健康和生态健康"三个维度，这三个维度层层递进又相互有机联系。其中个人健康包括生理健康和心理健康两个层次，即生命的活力、情绪的稳定和积极乐观的精神；社会健康是指个体的人与群体的人之间的关系处于良好的状态——和谐社会；生态健康则是指人类社会与自然环境之间的关系处于良好状态——可持续发展（郑丽，2020）。

According to The World Health Organization's definition of health, being healthy is not merely the absence of disease but a perfect state of physical, mental and social adaptability. In her book, *Therapeutic Horticulture and Healing Garden*, Zheng Li has made an extensive discussion on the conceptual extension of health. She believes that health can include three dimensions—"personal health, social health and ecological health". Among them, personal health includes the two levels of physical health and mental health, namely, the vitality of life, the emotional stability, and the optimistic spirit; social health refers to the good state of the relationship between a person as an individual and a person belonging to a group—a harmonious society; ecological health refers to the good state of the relationship between the human society and the natural environment——sustainable development (Zheng Li, 2020).

中国古代哲学思想认为，人内在的精神与自然的规律是密切联系的。人的精神对身心的健康是可以起到主宰作用的。《黄帝内经》曰："心者，君主之官也，神明出焉。"＊它把人的脏腑分为君臣关系，心是君，其他脏腑皆为臣。认为人的身体调节功能以心为主宰，而心则和情志、意识、精神等相关。换言之，就是情绪能调节人体的身心健康。所以，中医认为疾病是由于外感六淫和内伤七情而发生的。前者是外因，后者是内因，外因要通过内因起作用。六淫是指自然条件的变化，即风、寒、暑、湿、燥、火；七情就是指情绪的变化，即喜、怒、忧、思、悲、恐、惊。这是传统中医的理论，同时现代医学也有充分的临床资料和试验证据表明，情绪活动可以通过影响神经系统、内分泌系统和免疫系统的生理功能而导致疾病的发生。

According to the ancient Chinese philosophical thinking, the inner spirit of a person is closely related to the laws of nature, and the spirit can play a dominant role in physical and mental health. According to the *Inner Canon of the Yellow Emperor*, "the heart occupies the office of the Sovereign. Spirit luminosity stems from it."＊ The book draws an analogy between the human viscera and the relationship of the monarch and his ministers: the heart is the monarch, and the other viscera are all ministers. It is believed that the body's regulatory function is dominated by the heart, and that the heart is related to emotions, consciousness, and the spirit. In other words, emotions can regulate the physical and mental health. Therefore, in Chinese medicine, it is believed that a disease occurs due to the excessive or untimely working of the six natural factors and the internal injuries caused by seven emotions. The former is an external cause, and the latter an internal one. The external cause works through the internal one. The six natural factors refer to changes in natural conditions, that is, wind, cold, heat, damp, dryness, and fire; the seven emotions refer to changes in emotions, namely joy, anger, worry, thought, sadness, fear, and shock. This is the theory of traditional Chinese medicine. At the same time, modern medicine has sufficient clinical data and experimental evidence to show that emotional activities can cause diseases by affecting the physiological functions of the nervous, endocrine and immune systems.

3.1.2 基于花卉文化的花木精神与健康的关系

3.1.2 The relationship between health and the floral spirit based on flower culture

人类自古就了解花卉和园林环境对人体身心的调节作用，中国人通过养花赏花修身养性的传统也见证了花园对人类生活的积极作用。现代科学进一步证明，绿色景观能有效促进人们专注力的提升、调整情绪、改善身体机能；可见，园林在人类文明的发展史上贡献卓著。中国传统花文化就是一部博大精深的中国人修身养性的百科全

＊ 出自《黄帝内经·素问篇》——"灵兰秘典论"。

From *the Iinner Canon of Yellow Emperor · Su Wen*—"Ling Lan Mi Dian Lun".

书，不管是植物本身抑或植物构建的环境，都是人类健康照护系统中的重要因素。

Human beings have understood the regulatory effect of flowers and plants and the garden environment on mind and body since ancient times. Chinese people have also witnessed the positive effect of gardens on the human life through the tradition of growing and admiring flowers. Modern science further proves that green landscapes can effectively promote the improvement of concentration, adjust emotions, and improve somatic functions. This shows that gardens have made outstanding contributions to the development of human civilization. The traditional Chinese flower culture is a broad and profound encyclopedia of Chinese people cultivating moral characters. Whether it be plants themselves or the environment built by the plants, both are the important factors of the human health care system.

花的开谢是美学亦是哲学。花之美，在于其能牵动人对生命的细微感受。东吴大学物理学教授陈国镇先生在《情绪的"处方"：花精》一书的序言中说道：花朵不仅努力活出自己的特质，也无私地感染着周遭一切生物的情志。当我们接触到花时，自己的情绪、心理或心灵上的困扰自然被转化，因此，花对人而言，成为辅助生命灵性升华不可或缺的资粮。这种优雅中蕴藏的神奇力量，正是园林植物作用于人身心健康的基础（崔玖，2012）。花具有实体美的客观属性，又具有被人类主观审美的特质，是美的对象和象征物，是人们超越现实的精神内化物。大家都认为花是美的，赏花可以让身心愉悦。花是美的化身，美以真与善为条件而存在。爱美是人类的天性，如果爱美而不追求真与善，只能触及美的外壳，达不到爱美赏花之目的。赏花就是以花之美叩开心扉；以爱美之力，扫除虚假与邪恶、净化心灵，使精神世界开阔、乐观，少有烦恼与苦闷，从而保证良好的情绪、心灵的健康（徐海滨，1996）。中国花文化在一定程度上承载着中华民族的精神记忆，体现着中华民族的认同感、归属感。我国有丰富的花卉资源，如原产我国的牡丹、梅花、菊花、山茶等传统名花在中国园林发展史上都扮演着重要角色，备受人们喜爱，历代文人雅士对它们赞誉有加，在中式园林中的配置应用也表达了人们丰富的思想感情和文化特征。我国园林文化学家曹林娣先生说，植物蕴含着中国文化精神，规范着人们的审美创造；花卉丰富的艺术形态，也反过来改造和陶冶着人们的心灵。花木移情，若没有花木精神，便无所谓园林意境（曹林娣，2021）。

The blooming and fading of flowers is both aesthetics and philosophy. The beauty of flowers lies in its ability to affect people's subtle feelings about life! Soochow University's physics professor Chen Guozhen said in the preface of the book *Emotional "Prescription": Flower Essence* that flowers not only strive to live out a character of their own, but also selflessly produce an effect on the emotions of all the creatures around them. When we come into contact with flowers, our emotional, psychological or spiritual troubles are naturally transformed. Therefore, for people, flowers become an indispensable resource for the spiritual sublimation of life. The magical power contained in this elegance is the basis for the effect of garden flowers and trees on the physical and mental health (Cui Jiu, 2012). Flowers have the objective attribute of tangible beauty and the characteristic of being subjectively appreciated by humans. They are the object and symbol of beauty and people's internalized spiritual substance beyond reality. Most people are aware that flowers are beautiful, and by viewing them physical and mental pleasure can be generated. Flowers are the incarnation of beauty, and beauty exists on the condition of truth and goodness. The love of beauty is a human gift. If you love beauty without pursuing truth and goodness, you can only touch the outer shell of beauty and fail to achieve the purpose of loving beauty and admiring flowers. Appreciating flowers is to use the beauty of flowers to open one's heart and to use the power of loving beauty to wipe out falseness and evilness, purify one's soul, make the spiritual world open and optimistic and reduce trouble and depression to ensure good emotions and spiritual health (Xu Haibin, 1996). To a certain extent, the Chinese flower culture carries the spiritual memory of the Chinese nation, and embodies the Chinese nation's sense of identity and belonging. China has rich flower resources, and the traditional famous flowers such as the peonies, plum blossoms, chrysanthemums, and camellias native to the country, have played an important role in the development of Chinese gardens and been loved by the people and praised by refined scholars of past dynasties. Apart from that, the arrangement and application of those flowers in gardens have shown people's rich thoughts and feelings, as well as cultural characteristics. Cao Lindi, a well-known Chinese scholar of garden culture, believes that plants contain the spirit of Chinese culture and regulate people's aesthetic creation; The rich artistic form of flowers, in turn, transforms and edifies people's minds. Emotions are conveyed by flowers and trees. Without flowers and trees, there is no artistic conception of garden (Cao Lindi, 2021).

3.1.3 中式园林植物配置与身心健康的关系
3.1.3 The relationship between the plants arrangement of Chinese gardens and the physical and mental health

追溯历史，我们基于传统花卉文化表达的中式园林花木配置从以下几个方面彰显了植物环境与健康的关系，值得今人借鉴。

In retrospect, the flowers and trees arrangement of Chinese gardens based on the traditional flower culture has demonstrated the relationship between the plant environment and health from the following aspects, which is worth learning for people today.

（1）诗画山水的构图
（1）The layout of poetic and picturesque landscapes
中式古典园林在构图上很大程度受诗意山水理念的影响，常运用传统的诗词歌赋作为园林造景依据，并以诗文为景点题名。例如，苏州园林以植物命名的景点，约占总景点的四分之一，而"栽花种草全凭诗格取材""静赏有诗情，坐观有画意"，以诗词为依据，从历史上积淀的审美经验中，借鉴诗情画意的优美意境，营造"虽由人作，宛自天开"的特色，力求艺术地再现大自然之美，达到"天人合一"的境界。如网师园的"竹外一枝轩"* 取自苏轼的"江头千树春欲暗，竹外一枝斜更好"；狮子林的"暗香疏影楼"、拙政园的"雪香云蔚亭"等均源自赞梅的诗句。

The layout of Chinese classical gardens, in pursuit of the beautiful artistic conception of poetic and pictorial splendor, is largely influenced by the concept of poetic landscapes with the traditional poetry and verses often used as the basis for garden landscaping and poetic proses used to name scenic spots. For example, in the Suzhou gardens, the scenic spots named after plants account for about one quarter. "Planting flowers and growing grass is entirely guided by poetic standards". "There is poetic charm and artistic splendor in its appreciation". Grounded in historical aesthetic experiences, they draw inspiration from the beautiful connotations of poetry and painting. With the characteristic of "being artificial yet looking natural", Chinese classical gardens strive to reproduce the beauty of nature artistically and achieve "the harmony between man and nature"; for example, "The Prunus Mume Pavilion"* in the Master-of-Nets Garden in Suzhou is taken from Su Shi's poem: "The jungle on the riverside looks gloomy in spring, and a branch of plum blossoms extending out from the bamboo forest is even better." Apart from that, "The Slight Plum Sweet and Scare Shadow Tower" in Lion Forest Garden, and "The Snow-Like Fragrant Prunus Mume Pavilion" in Humble Administrator's Garden are derived from verses in praise of plum blossoms.

（2）比德叙事的立意

（2）The intention of morally uplifting narrative

除了优美的意境构图以外，中式古典园林在造园思想上也颇有讲究。每一座园林都是园主人精心营造用以修身养性的身心港湾，其造园的立意主题要么表达园主人的崇高理想，要么叙述园主人的情感际遇，或者是对美好生活的期盼。而在园林构建的所有要素中，借助植物的配置来比德叙事是最为妥帖和常见的。最典型的就是"四君子"梅兰竹菊、"岁寒三友"松竹梅、出淤泥不染之"青荷"、劲骨刚心之"牡丹"等。另外，人们还习惯以"桑梓"代表故乡，"椿萱"代表父母，"棠棣"代表兄弟，"兰草""桂树"代表子孙等，这些植物配置都表明古人的生存环境与植物亲切融合，寄托着园主人的情感和理想。

In addition to the beautiful layout for the artistic conception, particular care is also devoted to gardening ideas of Chinese classical gardens. Every garden is a physical and mental "harbor" meticulously created by the owner for self-cultivation. The central theme of the garden either expresses the owner's lofty ideal, narrates the owner's emotional experiences, or shows the owner's expectation of a wonderful life. Among all the elements of garden construction, morally uplifting narrative with the help of the layout of plants is the most appropriate and common. The most typical ones are the "Four Gentlemen"——plum,

---

* 景点英文名均出自《苏州古典园林：汉英对照》。

All scenic spot names are taken from *Classical Gardens of Suzhou：Chinese-English Comparison*.

orchid, bamboo and chrysanthemum, the "Three Friends of Winter"——pine, bamboo and plum, the "Green Lotus" that does not stain the mud, and the "Peony" that is strong in bone and heart. In addition, people are accustomed to using "Sangzi" mulberry tree and catalpa tree to represent their hometown, "Chunxuan" the Paulownia and the Forget-me-not to represent their parents, "Tangdi" to represent their brothers, "Orchid Grass" and "Osmanthus Tree" to represent their descendants, etc. These plant arrangements all indicate the friendly integration of ancient living environment with plants, embodying the emotions and thoughts of the garden owner.

（3）四季更替的表现
（3）The reflection of alterations of four seasons

中式园林说到底，承载人居环境的功能。园主人在此休养生息、居家度日，这势必成为人与自然最为直接的一个联系载体。童寯先生在《江南园林志》中写道："园林无花木则无生气。盖四时之景不同，欣赏游观，怡情育物，多有赖于东篱庭砌，三径盆盎，俾自春迄冬，常有不谢之花也。"欧阳修也曾作诗："浅深红白宜相间，先后仍须次第栽。我欲四时携酒去，莫叫一日不花开。"自然的物候变化可经由植物的表现而让人有所知觉。这也可看作是人、植物、天地之间的一种默契。因此，让人修身养性的园林就教会了人们对时令的尊重，结合春夏秋冬季节更替与生命春生夏长秋收冬藏的规律，建立人与植物、人与自然的和谐关系。

Chinese gardens function as a living environment. The owner of a garden recuperates, builds up strength and lives a family life there, which certainly makes the garden the most direct medium for contact between man and nature. Mr. Tong Jun wrote in the *Jiangnan Garden Chronicle* that "a garden without flowers and trees is lifeless. The scenery of the four seasons is different, and appreciating the scenery and nurturing things with joy often relies on the construction of the eastern fence and three paths of pots full of flowers, so that from spring to winter, there are often flowers that do not wither." Ouyang Xiu also wrote a poem, "Light, deep, red, and white should be mixed, and they still need to be planted one after another. I want to appreciate flowers with wine in any season, not hoping that there is a single day without blossoms." Nature's phenological changes can be perceived through the performance of plants. This can also be seen as a tacit understanding among the humankind, plants, and the universe. Therefore, a garden that enables self-cultivation shows people's respect for the seasons. By combining the shift of spring, summer, autumn and winter with the laws of sowing in spring, developing in summer, harvesting in autumn, storing in winter, the garden establishes a harmonious relationship between humanity and plants, and between man and nature.

（4）生活物资的提供
（4）The provision of living materials

植物对人类生活的贡献众所周知，人们的衣食住行无一能离开植物。中式园林中的植物也承担着照护园主人生活所需的功能。中国传统的花文化体系涵盖了人类在长期和自然交流的过程中，总结出的一套园林植物栽培应用的方法。包括如何培养植物，如何让植物经加工变成食物，后来又在物质食物的基础上演变出精神食粮——花卉的欣赏、园林的营造。例如，苏州的私家园林就带有一定的自然经济色彩，能园产自给。植物的根、叶、花、果，不仅可作四时清供，实际上还能产生经济效益。从拙政园取名依据的诗文"筑室种树，逍遥自得……灌园鬻蔬，以供朝夕之膳"，可见园林植物最初服务于生活的功用。宋代朱长文乐圃"药录所收，雅记所名，得之不为不多。桑柘可蚕，麻苎可绩，时果分蹊，嘉蔬满畦，摽梅沉李，剥瓜断壶，以娱宾客，以酌亲属"的记载就是园林提供生活物资的经典描述。除此之外，梅宣义的五亩园、范成大的石湖别墅、韩雍的蓊溪草堂等，也都具有很强的田园生产功能。

The contribution of plants to human life is well known, and food, clothing, shelter and means of traveling——the four basic needs of everybody cannot go without plants. The plants in Chinese gardens undertake the function to meet the living needs of garden owners. The traditional Chinese flower culture system involves a set of methods to cultivate and use garden plants summarized by humankind during the long-term interaction with nature, including ways to cultivate plants and to process plants into food, and later, the spiritual food——the appreciation of flowers and the construction of gardens——was derived from physical food. For example, private gardens in Suzhou have certain characteristics of natural economy, allowing for self-sufficiency in garden production. The roots, leaves, flowers, and fruits of plants can not only be used for seasonal display, but also generate economic benefits. The naming of the Humble Administrator's Garden is named after the poem "Building a house and planting trees, enjoying oneself freely... Irrigating the garden and selling vegetables for daily meals", which shows the original function of garden plants in serving daily life. The record of Zhu Changwen's Lepu in the Song Dynasty, "The collection of medicinal records and the names of elegant records are quite a few. Mulberry can be used for silkworms, hemp can be used for spinning, seasonal fruits are plenty, and the fields are filled with fine vegetables. Plums are marked, and melons and gourds are peeled to entertain guests and drink with relatives", is a classic description of the garden providing daily necessities. In addition, Mei Xuanyi's Five-acre Garden, Fan Chengda's Shihu Villa, and Han Yong's Fengxi Thatched Cottage all have strong pastoral production functions.

可见，中式园林通过植物配置对园主人身心健康产生积极作用由来已久。中国古典园林的不少名园主人，虽然在仕途官场遭受人生挫折、历经心灵创伤，所幸在自家的花园里找到了身心的慰藉，获得了自我疗愈。例如，扬州何园主人何芷舠，49 岁时因看透官场的昏暗，卸任到扬州，购得吴氏片石山房旧址，后扩为园林建造"寄啸山庄"，后称何园，园名取自陶渊明的《归去来兮》"依南窗以寄傲，登东皋以舒啸"。他在这里侍奉老母，年近七旬又重出江湖，征战商场，创下不朽业绩。不难说是寄啸山庄让他养精蓄锐，给了他过人精力。

It can be seen that Chinese gardens have a long history of positive effects on the physical and mental health of garden owners through plant configuration. Although many owners of famous Chinese classical gardens suffered life setbacks and experienced psychic traumas in official careers, they fortunately found physical and mental comfort in their gardens and thus healed themselves. For example, when He Zhidao, the owner of He Garden, was 49 years old, he left his official post as governor in Hubei for Yangzhou after thoroughly understanding the darkness of the officialdom, and then he bought a Wu family's former site of the Pianshishanfang Garden and expanded it into a garden named "Jixiao Villa", which then was renamed as He Garden. The name of "Jixiao Villa" is taken from a verse in *Returning Home*, a poem written by Tao Yuanming: "I lean upon the southern window with an immense satisfaction, or go up to Tungkao and make a long-drawn call on top of the hill." He Zhidao served his mother there, and returned to the scene, conquered the business world in his late sixties and made an immortal achievement. It stands to reason that it was the Jixiao Villa that had given him the extraordinary energy by enabling him to nourish spirit and store up vigor.

3.1.4　花木精神力量的来源浅析
3.1.4　A simple analysis on the source of the spiritual power of flowers and trees

综上所述，我们将对花木精神力量的来源做一尝试性分析。植物是园林中唯一具有生命力的构成要素，如苏州园林就讲究一年无日不看花。根据花木的季相特点，设四季之景，如拙政园的海棠春坞、荷风四面亭、待霜亭、雪香云蔚亭就是四季景点的代表；花木品种的选择和配置，需在科学的前提下做到"赏心悦目"，重在精神情感上的享受。根据花木的生态习性、色彩、同音或谐音，往往被人们赋予特定的象征意义和情感寄托。如园林中较多种植金银花和枇杷，均色黄如金，是吉利的象征；中式庭院象征"金玉满堂"的'金桂'、玉兰、石榴和海棠也都是常见树种；中国传统十大名花的人文精神象征意义更是家喻户晓。因此，抛开植物提供给人类生活物资方面的贡献，单从植物文化中所表达的蕴含于花木中的精神力量来看，其对人的思想感情和身心健康的影响或许来源于以下几方面：

Considering the above, we will conduct a tentative analysis of the sources of the spiritual power of flowers and trees. Plants are the only vital element in gardens, such as Suzhou gardens, which emphasize the importance of looking at flowers every day of the year. Based on the seasonal characteristics of flowers and trees, we set up scenic spots of the four seasons, such as "The Malus Spectabilis Garden Court" "The Pavilion in Lotus Breezes" "The Orange Pavilion" and "The Snow-Like Fragrant Prunus Mume Pavilion" in Humble Administrator's Garden, which are representative of seasonal scenic spots. The selection and configuration of flower and tree varieties should be based on scientific principles, with a focus on enjoying spiritual and emotional pleasure. According to the ecological habits, colors and homophones of flowers and trees, they are often endowed with specific symbolic meanings and emotional sustenance by people. For example, honeysuckle and loquat are planted more in the gardens, with a uniform yellow color as gold, which is a symbol of good luck.

Osmanthus fragrans in Chinese courtyards, symbolizing the abundance of gold and jade, yulan magnolia, pomegranate and crabapple are also common tree species, let alone the symbolic significance of the humanistic spirit of Chinese Top Ten Traditional Famous Flowers. Aside from plants' contribution in providing living materials in the fields such as clothing, food, housing, and transportation, even from the perspective of the spiritual power contained in flowers and trees expressed in the plant culture, plants' influence on people's thoughts, feelings, and physical and mental health may come from the following aspects:

(1) 亲自然的归属感
(1) A sense of belonging to nature

人类的生存自古就离不开植物，因此植物能带给人安全感和归属感，人可以因为植物的存在而感到生活的保障、心灵的安定和情绪的平静。

Human beings have not been able to live without plants since ancient times. Therefore, plants can bring people a sense of security and belonging; and due to the existence of plants, people can have a sense of living security, the peace of mind and the calm of emotions.

(2) 穿越历史的神圣感
(2) A sense of sacredness through the ages

人类对大自然的敬畏之心很多时候是通过植物来作为表达载体的，比如祭祀、图腾等。面对一棵百年老树，如仰头凝视网师园濯缨水阁北侧逾百年树龄的白皮松延展至水面上空的树冠(图3-1)，人会发自内心地产生崇拜和感慨，肃然而生敬畏。而这

图3-1　网师园白皮松延展至水面上空的树冠与濯缨水阁交相辉映(郑丽　摄)
Fig. 3-1　In the Master-of-Nets Garden, the Lacebark Pine's canopy stretches gracefully, echoing the elegance of the Washing My Ribbon Pavilion over the Water (by Zheng Li)

种对自然的敬畏有如内心的精神支柱，能让人保持身心的神明安稳，如端坐于苏州大学校园内逾百年树龄的鸡爪槭树下，内心不自觉就获得宁静安详(图3-2)。

Human reverence for nature is often expressed through plants, such as sacrifices and totems. In the face of a century-old tree, people would worship and sigh with emotion from the bottom of their heart, and be struck with awe when gazing up at the canopy of a Lacebark Pine tree that has been stretching above the water surface for over a hundred years on the north side of the Washing My Ribbon Pavilion over the Water in the Master-of-Nets Garden (Fig. 3-1). And this kind of reverence for nature is like the inner spiritual support, which can keep the mind and body stable. For example, sitting under the centennial Japanese Maples on the campus of Suzhou University, one's heart unconsciously obtains peace and tranquility (Fig. 3-2).

图 3-2　苏州大学天赐庄校园内逾百年生鸡爪槭给人的宁静安详(郑丽　摄)

Fig. 3-2　The peace and tranquility of the centennial Japanese Maples on the Tiancizhuang Campus of Soochow University (by Zheng Li)

(3)生命力的崇高感
(3)A sense of sublimeness of vitality

植物对自然的适应能力远在人类之上，其顽强的生命力和旺盛的繁殖能力是人类未曾企及的。如许多芳香植物，生长在艰苦贫瘠的环境中，经历了苦难，却能够沉淀出生命的芬芳。所以不管是傲霜斗雪的梅花、不畏严寒贫瘠的苍松，还是耐得住高温潮湿的檀香，它们都能给人一种勇敢向上的精神力量，带着人类走出困境、渡过难关。

The adaptability of plants to nature is far higher than that of humankind, and the tenacious vitality and vigorous reproductive ability of flowers and trees are what humans do not have and hope for. For example, many aromatic plants grow in harsh and barren environments, experiencing hardships, but being able to precipitate the fragrance of life. Therefore, whether it be the cold resistant Plum Blossoms or the cold-and barren-resistant green pines, they can bring a brave and upbeat spiritual power into people's heart and lead humankind to get out of difficulties and to survive crises.

（4）多维度的美感

（4）The multi-dimensional beauty

植物的色、香、姿、韵可谓从外在美和内在美的维度满足了人类五感的审美需求，这种来自大自然的生命之美，是人类身心保持安宁祥和的最佳滋养。植物承载的生命信息和力量也因此注入人体，带领人类走向天人合一的境界。

It may be said that the color, fragrance, posture, and charm of plants, in terms of external and internal beauty, have met the aesthetic needs of humankind's five senses. The beauty of life, which is from nature, is the best nourishment for maintaining peace and harmony in the body and mind of human beings. The life information and power contained in plants are thus injected into the human body, leading humankind into the realm of the unity of nature and man.

综上所述，对园林植物的培育能促进身体与自然的亲和，在对植物的照护中得到自身锻炼以达个人的健康；赏花能改善人们的身心健康，促进社会的和谐稳定；种花植草美化环境，共建生态安全。

In summary, we can see that the cultivation of garden plants can promote the body's closeness to nature, and personal health can be achieved through the workouts in the care of flowers and plants; appreciating flowers can assist people's spiritual and emotional health, promote the harmony and stability of society; planting flowers and plants can beautify the environment and jointly achieve ecological security.

显然，园林作为创造与改善城市环境的主要手段，已成为促进公共健康的重要组成部分。不管是植物本身抑或植物构建的环境，都是人类健康照护系统中的重要因素。其实，这也是城市绿地空间于人类而言的初衷所在。因此，城市园林中的自然环境、活动空间必将成为市民改善自身健康状态的重要场所。而关于植物文化的精神力量对身心健康的作用也值得探索和实践。

Obviously, gardens, as the main means of creating and improving the urban environment, have become an important part of the promotion of public health. Both the plants or the environment built by them are important factors in the human health care system. In fact, this is the original intention of the urban green space for humankind. Therefore, the natural environment and activity space in urban gardens will surely become important venues for citizens to improve their health. It is worth exploring and giving play to the effect of the spiritual power of the plants culture on the physical and mental health.

3.2 现代景观康复的主要理论

3.2 The main healing theories of modern healing gardens

3.2.1 恢复性环境理论

3.2.1 The theory of restorative environments

恢复性环境（restorative environments）是环境心理学的前沿研究（苏谦，2010），其主要关注自然环境对人体心理状况的改善，包括对压力、情绪等的调节。恢复性环境以压力恢复理论（SRT）和注意力恢复理论（ART）为理论基础，解释了自然环境对人体压力与注意力的恢复机制。

Restorative environments are the cutting-edge research of environmental psychology (Su Qian, 2010), mainly focusing on the improvements of physical and psychological conditions by the natural environment, including the regulation of stress and emotions. The theory of restorative environments is based on the Stress Reduction Theory (SRT) and the Attention Restorative Theory (ART), which have explained the natural environment's restoring mechanism for human stress and attention.

（1）压力缓解理论

（1）The Stress Reduction Theory (SRT)

压力缓解理论由美国得克萨斯农工大学的罗杰斯·乌尔里希教授提出（Ulrich，1983），是指当个体处于压力状态下，会导致消极情绪、生理系统短期变化和异常行为等不良状况，而当人处于自然元素丰富、优美的环境下时，注意力容易发生迁移，从而阻断消极情绪的产生或以积极情绪代替消极情绪，带来与之相应的机体恢复。乌尔里希强调情绪是个体不需要由认知调节的、对环境刺激的首要和直接反应，对优美环境的偏好是与生俱来的倾向与积极反应，是长期进化的结果，该理论又称心理进化理论。

The SRT was proposed by Professor Roger Ulrich of Texas A&M University (Ulrich, 1983). He believes when an individual is under stress, the stress would lead to negative emotions, short-term changes in the physiological system and abnormal behavior, and that when a person is in a beautiful environment with rich natural elements, his or her attention is prone to shift, which thereby blocks the generation of negative emotions or replaces negative emotions with positive ones, bringing about the body recovery accordingly. Ulrich emphasized that the primary and direct response to environmental stimuli is emotions, which do not need to be regulated by cognition. In addition, the preference for beautiful environments is a tendency and positive reaction one already has without learning, a result of long-term evolution. The theory is also called the Theory of Psychological Evolution.

（2）注意力恢复理论

（2）The Attention Restorative Theory (ART)

注意力恢复理论由卡普兰夫妇基于自主性和非自主性注意力差别的概念提出（Kaplan，1989，1995）。他俩认为，人们为了日常生活的高效率必须保持认知清晰，清晰的认知需要"主动注意"来维持，若维持"主动注意"的能力不足，则会导致很多负面的影响，如应激性降低，没有能力做计划，对人际关系信息的敏感性降低，认知作业错误率上升等。人在维持主动注意的时候耗能较大，易于疲劳；而自然环境使人产生的"被动注意"则是一个耗能较低的注意模式，这种模式可以对主动注意消耗的能量进行恢复。例如，人们在花园中散步不经意间闻到的花香就是属于被动注意模式，该理论即解释了为何闻到花香人会感觉精神放松。

The ART was proposed by the Kaplan couple based on the concept of the difference between voluntary and involuntary attention（Kaplan, 1989, 1995）. They believe that in order to carry out daily life efficiently, people must maintain clear cognition, whose maintaining requires "directed attention". The decline in the ability to concentrate directed attention will lead to many negative effects, such as decreased sensitivity, inability to plan, decreased sensitivity to interpersonal information, and an increased error rate in cognitive tasks. When people concentrate directed attention, they consume a lot of energy and are susceptible to fatigue. But the "passive attention" that the natural environment makes people generate is an attention mode with lower energy consumption. Such a mode can restore the energy consumed by directed attention. An example of this is: the scent of flowers that people inadvertently smell when walking in a garden falls into the passive attentionmode. This theory can explain why people feel more relaxed when they smell the scent of flowers.

压力缓解理论和注意力恢复理论被视为当前康复景观研究领域最重要的两个理论，两者均试图从心理学角度对自然体验与人体健康之间的关系进行确切归因，均假设人们对自然环境有一种强烈且一致的积极趋向，即偏好。正如苏州沧浪亭所挂对联"清风明月本无价，近水远山皆有情"，道出了自然美景是大自然赏赐给人类的无价之宝，近处的水、远处的山都饱含深情，叫人如何能不趋之！

The SRT and ART are regarded as the two most important theories in the field of healing landscape research. Both attempt to find out the exact relationship between natural experiences and human health from a psychological perspective, and both assume that people have a strong and consistent positive tendency for the natural environment, that is, a preference. As the couplet hanging at the Canglang Pavilion in Suzhou goes, "clear breeze and bright moons are priceless, and both near waters and far mountains have emotions". It expresses that natural beauty is a priceless treasure bestowed upon humanity by nature. The nearby water and distant mountains are full of deep emotions, how can people not follow it！

3.2.2　环境偏好理论
3.2.2　The environmental preference theory

（1）环境偏好模型

卡普兰夫妇提出的环境偏好模型（Kaplan, 1989）指出，能够被人类喜欢或者偏好的景观，是那些在本质上适合人类，不单调、不乏味，并能够被人类所运用的环境；同时这种环境具有刺激人体信息加工的能力。由此，该模型提出一个包含四类组成的环境偏好矩阵（表3-1）：一致性、易读性、复杂性和神秘性。其中一致性和易读性使人能够理解环境，而复杂性与神秘性是激发人们探索的动机，可见这一矩阵是"理解"与"探索"的对立。对环境偏好而言，一致性与复杂性达到中等水平即可，而易读性与神秘性则越多越好，前一组体现的是推理分析，后一组体现的是认知加工。

Kaplan and his wife proposed the environmental preference model (Kaplan, 1989), believing that the landscapes liked or preferred by humans are those that are not monotonous and boring but essentially suitable for humans and can be used by humans; at the same time, this environment can stimulate the human body's ability to process information. Based on this, Kaplan and his wife proposed an environmental preference matrix consisting of four categories (Tab. 3-1): compatibility, legibility, complexity, and mystery. Among them, consistency and legibility enable people to understand the environment, while complexity and mystery are the motivations for people to explore. It stands to reason that this matrix is a contrast between "understanding" and "exploration". For environmental preferences, it is sufficient for consistency and complexity to reach a moderate level, while the more legible and mysterious it is, the better. Consistency and complexity embody rational analysis, while legibility and mystery embody cognitive processing.

表 3-1　环境偏好模型

Tab. 3-1　The environmental preference model

信息的特点 Characteristics of information	理解性 Comprehensibility	探索性 Explorativeness
直接的 Direct	一致性 Consistency	复杂性 Complexity
推断的和预测的 Deductive and predictive	易读性 Legibility	神秘性 Mystery

环境偏好模型向人们解释了——优美的自然环境为大多数人所偏好，是基于人类精神上对美的需求和渴望；而有趣的园林环境为人们所偏好，则是基于人类行为上的认知和探索需求。

The environmental preference model explains that the reason the beautiful natural environment is preferred by most people is based on the humankind's spiritual need and desire for beauty, while the reason the interesting garden environment is preferred by people is based on the needs, in terms of human behavior, to know and explore.

(2) 环境偏好特性

(2) Environmental preference characteristics

卡普兰夫妇通过对人们在园林中获得的满意度自评调研进行总结，还提出了以下思考，他们认为人造景观若要达到疗愈效果，需要具备抽离性、延展性、迷人性与一致性四个方面的特点。这可以归纳为康复花园设计所需具备的环境偏好特性。

The Kaplan couple summarized through a self-evaluation survey about the satisfaction people get from the garden, and put forward the following thought: They believe that for an artificial landscape to have a healing effect, there should be four key components: being away, extent, fascination, and compatibility. These are just environmental preference characteristics that are essential to healing garden design.

①抽离性 特指身体和心理上的抽离。身体上的抽离，比如暂时离开固有的生活、工作环境；心理上的抽离可以是任何使心思脱离现实生活的情绪体验。观察自然、置身花园中是一种具有普遍意义的抽离。仅仅从窗外看见缤纷的花朵，或者到花园里走一走也可以产生抽离感。

①Being away refers specifically to being away physically and psychologically. To be away physically, one can temporarily leave the old life and work environment, while to be away psychologically, one can have any emotional experience that makes the mind detached from real life. Observing nature and being in a garden are "being away" in a general sense. One can get a sense of being away just by seeing the colorful flowers from the window or walking in a garden.

②延展性 指人们所处的环境拥有充足的内容和结构，在时间与空间上具备有形的或无形的扩展。这样的环境能够占据人们大脑一段足够长的时间，以使人能从主动注意的集中模式状态得到休息和放松。

②Extent refers to the sufficient content and structure, with tangible or intangible extension in time and space, boasted by the environment that people are in. Such an environment can occupy people's mind for a long enough time so that they can get out of the concentration mode of directed attention to rest and relax.

③迷人性 指环境本身充满了吸引力，让人不需要去消耗大量的能量就能专注其间，即转化为能耗较低的被动注意模式，同时对主动注意消耗的能量具有恢复和补充的作用。自然环境中如云、夕阳、风景、微风中树叶的摆动等，这些可读的信息经常以一种不具有刺激性的魅力吸引着人们的注意，使人身心放松，这对人的健康具有裨益。

③Fascination means that the environment itself is full of attractiveness, so that people can focus on it without consuming a lot of energy. In other words, people can shift to a passive attention mode with lower energy consumption. And at the same time, such an environment can restore and replenish the energy consumed by directed attention. Clouds, sunsets, landscapes, the leaves swaying in the breeze, and so on in natural environments, whose readable messages often attract people's attention with a non-irritating charm, helping people relax both physically and mentally, which is beneficial to human health.

④一致性 指使用者游园的目标或倾向与经由环境所提供的、引导的或要求的活动内容有较高的契合度。环境类型要满足人们的爱好与行为需求。人们对自然有几种认识：作为捕食者，如人可以在自然中狩猎、打鱼；作为运动者，如人可以远足、划船；作为驯化者，如可以照顾花草、饲养动物；作为观察者，如观察鸟儿、参观动物园；作为生存者，如生火、搭建庇护设施等。以上这些情景，在人们接触自然之前已经在头脑中有所认知，这使得环境与人之间更具备了一致性。

④Compatibility refers to the high-degree compatibility between the user's goals or inclination during the visit to a garden and the activities provided, guided or required by the environment. The type of the environment should satisfy people's hobbies and behavioral needs. People have several understandings of nature: as a predator, people can do things like hunting and fishing in nature; as a sportsman, people hiking and rowing a boat; as a domesticator, people taking care of flowers and plants and raising animals; as an observer, people observing birds and visiting zoos; as a survivor, people lighting fires and establishing shelter facilities. The above scenarios have been recognized in people's minds before they engage with nature, which makes the environment and people more compatible.

3.3　植物通过五官感觉对健康的作用
3.3　The effect of plants on health through the five senses

植物作为大自然生态圈最重要的构成要素之一，对人体健康的积极作用是多层次、全方位的。本节主要从视觉、听觉、嗅觉、味觉和触觉层面，就植物如何通过五感对人体健康发挥作用进行阐述。同时，五感滋养体验也是园艺康健中较重要的部分，而关于园艺康健将在第4章进行概述。

As one of the most important components of the natural ecosystem, plants have a multi-level and all-around positive effect on human health. This section mainly discusses how plants play a role in human health through the five senses from the perspective of sight, hearing, smell, taste and touch. Meanwhile, the experience of nourishing the five senses is also an important part of horticultural healing which will be summarized in Chapter 4.

3.3.1　视觉
3.3.1　Sight

视觉是光线刺激眼睛视网膜上的感光细胞，经视觉神经系统加工后在人脑中形成的关于形状、大小、颜色、动静等外界物体特征的综合信息。据不完全统计，人类有约80%的外界信息是通过视觉获得的，因而其对人类认知乃至思维能力等的发展至关重要。

Sight is the comprehensive information, processed by the visual nervous system formed in the human brain after the light stimulates the photoreceptor cells on the retina of eyes, about characteristics of external objects, such as the shape, size, color, and movements. According to incomplete statistics, about 80% of the external information of human beings is obtained through sight, so sight is very important for the development of humankind's cognition and even thinking ability.

随着学界对人类视觉机制研究的不断深入，视觉环境与人类情绪状态、睡眠质量，以及工作效率等方面的关系被逐步揭示，通过改善视觉环境促进人类健康已成为设计行业发展的重要方向之一。植物因为色彩丰富，姿态各异，极具观赏价值，因而是视觉环境设计的重点之一。伴随春夏秋冬的季节轮回，植物会呈现盛衰枯荣等不同生命状态，还会汇集大量昆虫、啼鸟，引来蝶舞、蜂忙，这些是环境设计中最富生机的视觉要素。不同颜色可以提供不同的视觉效果。暖色如红色、橙色、黄色等较为鲜艳夺目，使人心跳加速、精神亢奋，给人以热烈、辉煌、兴奋和温暖的感觉；冷色如青色、蓝色、紫色等较为深沉，则使人感到清爽、娴雅、肃穆、宁静和放松；白色花卉令人感到神圣纯洁和宁静，具有消暑的作用。利用园艺植物的颜色可以进行治疗。试验证明，浅蓝色的鲜花对于高烧患者具有良好的镇静作用；紫色的鲜花可使孕妇心情愉快；红色的鲜花能增进患者的食欲及增强听力；赭色的鲜花对低血糖患者大有裨益；绿色的花叶能吸收阳光中的紫外线，减少对眼睛的刺激，对眼睛有保护作用，并能消除心神疲劳。

With the academic circle's continuous in-depth study of the human visual mechanism, the relationship between the visual environment and the human emotional state, sleep quality, work efficiency and so on has been gradually revealed. Promoting human health by improving the visual environment has become one of the important directions for the development of the design industry. Plants are of great viewing value because of their rich colors and different postures, thus becoming one of the key points of the visual environment design. Along with the seasonal change of spring, summer, autumn and winter, plants present different life stages such as prosperity and decline, bloom and decay. Plants also bring together many insects and songbirds and attract dancing butterflies and busy bees, which are the most vibrant visual elements in an environmental design. Different colors can deliver different visual effects. Warm colors such as red, orange, yellow, and so on, are more resplendent, making people's heart beat faster, cheering people up and giving them a feeling of enthusiasm, brilliance, excitement and warmth; cool colors such as cyan, blue, purple, etc. are relatively more subdued, giving people a feeling of refreshment, elegance, solemnness, tranquilness and relaxation; white flowers give people a sense of sacredness, pureness and tranquilness, and have the effect of relieving summer heat. And consequently, we can use colors of horticultural plants for treatment. Experiments have proved that light blue flowers have a good calming effect on patients with a high fever; purple flowers can cheer pregnant women up; red flowers can increase the appetite of patients and enhance hearing; ocher flowers are of great benefit to patients with hypoglycemia; green flowers and leaves can absorb ultraviolet rays from the sun and reduce the irritation to eyes, so they have a protective effect on eyes and can relieve mental fatigue.

美国学者罗杰·乌尔里希对宾夕法尼亚州一所城郊医院胆囊手术患者的术后康复研究，可谓植物景观促进人体健康的权威研究（Ulrich，1984），在该研究中，患者分成 2 组：一组从病房只能看到一面砖墙；另一组从病房可以看到一片小树林（图 3-3）。分组依据性别、年龄、吸烟与否、肥胖或正常体重、住院经历以及手术年度等平均分配，包括 15 对女性患者与 8 对男性患者，随机安置到同楼层一样的病房里，只有窗景各异。虽然他们的手术医生不同，但均由同样的护理人员看护。

American scholar Roger Ulrich's postoperative recovery study on patients having undergone a gallbladder surgery in a suburban hospital in Pennsylvania can be described as an authoritative study on plant landscapes' promoting human health (Ulrich, 1984). In this study, patients were divided into two groups: one group was put in the wards with a view of a brick wall, while the other group a view of a small forest (Fig. 3-3). According to gender, age, smoking or not, obesity or normal weight, hospitalization experiences, and the year of operation, the groups were divided equally, including 15 pairs of female patients and 8 pairs of male patients, who were randomly placed in the same wards on the same floor, with different window views. Although their surgeons were different, they were all looked after by the same nursing staff.

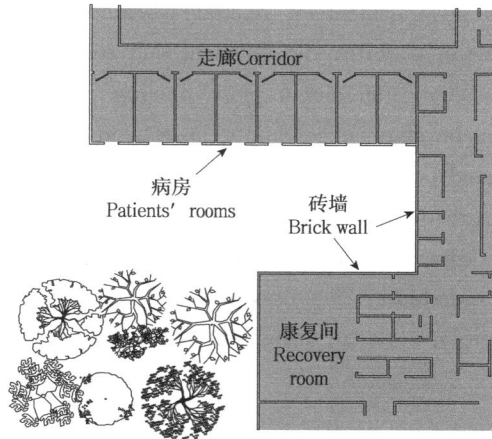

图 3-3 病房示意图
Fig. 3-3 Schematic diagram of wards

　　乌尔里希让一位具有丰富外科护理经验的护士详细检查了这 46 名患者的记录，她并不知晓患者所入住病房的窗景差异。这位护士着重检查三组基础数据：①患者住院期间消耗的强止痛药数量；②患者向护士抱怨的次数；③患者康复出院的时间。记录明确显示：23 位入住小树林窗景病房的患者所消耗的强止痛药数量较小，抱怨较少，且较入住砖墙窗景病房的 23 位患者住院时间短，其医疗费用平均少花 500 美金。该研究开启了大自然对人类健康积极作用的循证研究之先河，推动了具有健康效应的康复空间植物景观循证设计。

Ulrich asked a nurse with rich nursing experience on surgical floors to check the records of those 46 patients in detail. This nurse was not briefed on the different window views of different patients' wards. The nurse focused on three sets of basic data: ①the number of strong painkillers consumed by the patient during hospitalization; ②the number of complaints by the patient to the nurse; ③the time the patient recovered and was discharged from the hospital. The records clearly showed that the 23 patients who stayed in a ward with a window view of a small forest consumed fewer strong painkillers and complained less, spending USD 500 less on average in medical expenses, and their hospital stay was shorter than the 23 patients who stayed in a ward with a window view of a brick wall. This was the first evidence-based research on the positive effects of nature on human health, and promoted the evidence-based design of the plant landscapes of the healing spaces with health care effects.

随着相关研究的不断推进，植物对促进人类认知能力发育，降低压力水平等方面的功效日益被接受，并发展出针对某些特殊患者群体的、以植物观察为核心的专业性植物疗愈项目。如许多特殊学校都将植物观察纳入日常教学内容。事实证明，这些观察活动对促进儿童智力发育，尤其是促进认知功能障碍儿童的身心健康十分有益。这些研究为提升植物景观视觉设计的疗愈功能提供了科学依据，促进了植物景观的视觉设计向更加健康的方向发展。

With the continuous advancement of related researches, the effects of plants on promoting humankind's cognitive ability development and reducing stress levels have been increasingly accepted, and the professional plant healing projects targeting some special patient groups, with a focus on plants observation, have been developed. For example, many special schools have included plants observation into their daily teaching content. And facts have proven that these observation activities are very beneficial to promoting the intellectual development of children, especially to the physical and mental health of children with cognitive dysfunctions. These studies have provided scientific bases for improving the healing function of the plant landscape visual design, and promoted its development towards a healthier direction.

3.3.2 听觉
3.3.2 Hearing

听觉是声波所引发的人耳鼓膜振动经听小骨放大之后传到内耳，刺激耳蜗内的纤毛细胞产生神经冲动，再经听神经传到大脑皮质的听觉中枢，通过各级听觉中枢分析后形成的感觉。

Hearing is a sense formed after the auditory centers at all levels analyze the sound waves that cause the tympanic membrane vibration, which is amplified by the auditory ossicles and then transmitted to the inner ear to stimulate the ciliary cells in the cochlea to produce nerve impulses, which are then transmitted to the auditory centers of the cerebral cortex through auditory nerves.

人类很早就懂得"听"可以陶冶性情，对促进人体的身心健康有积极作用。中国古代最伟大的思想家、教育家孔子就十分主张"乐教"，调音弄曲也一直是中国传统士大夫主要的养生手段之一。随着现代科学的发展，人们进一步掌握了音乐疗愈各种疾病的机制，尤其在治疗各种心理疾病方面，有其他疗愈手段无法企及的优势。美国于 20 世纪 80 年代推出音乐疗法，目前已成为替代医学的重要组成部分，音乐治疗师也成为在美国拥有专业医疗资质的职业门类之一。

Human beings have long understood that one can cultivate temperament by "hearing", which has a positive effect on promoting the physical and mental health. Confucius, the greatest thinker and educator in ancient China, strongly advocated "music education", and tuning tones and playing music has always been one of the main methods of health preservation for traditional Chinese scholar-officials. With the development of modern science, people have further grasped the mechanism of music healing various diseases, and other healing methods cannot match music in terms of advantages, especially in the treatment of various mental diseases. Music therapy was introduced in the United States in the 1980s and has now become an important part of alternative medicine, and music therapists have become one of the professional categories that have professional medical qualifications in the United States.

植物本身并不会发出声音，但植物能够汇集各类昆虫、鸣禽，为园林带来蝉噪鸟鸣的自然声环境。此外，大自然风吹雨打等气候现象经植物的媒介作用，可为花园带来各种各样的声响，不仅动听，还具有很好的康健效益。落叶随风发出的瑟瑟声、青草摇曳的沙沙声、雨打芭蕉的滴答声、小鸟的叫声、花园里的风声，均能制造出不同的听觉效果，产生听觉刺激，让人感受大自然的美妙和转变。中国古代把风吹竹林的声音比作"戛玉"，形容这种声音像敲击碎玉石那么好听。清代名医周之缘《医心集·竹石篇》曰："听竹清心，伴竹得乐，尤胜于服药者也。"足见古人对声音疗愈功能的认识由来已久。同时，树木、篱笆、灌木丛可以隔离噪声，为人们提供宁静松弛的空间，有助于思考和冥想。也可以通过花园中安装的风铃或雨铃，增加听觉刺激效果。这种通过"零设计"的处理手法可以使人感受真实的自然声音，以达到心境的松弛平和。

Plants do not make sounds, but they can draw together all kinds of insects and songbirds, and create an environment with natural sounds like chirps of cicadas and birds; in addition, wind, rain and other climatic phenomena of nature can bring a variety of sounds, not only beautiful but also beneficial to health, to the garden through the media of plants. The rustle of fallen leaves in the wind and grass swaying, the pitter-patter of rain dropping on plantains, the chirps of birds, and the sound of wind in the garden can all create different auditory effects, produce auditory stimulation, and make people feel the beauty and changes of nature. In ancient China, the sound of the wind blowing in the bamboo forest was likened to "jade clanking" to describe the sound as pleasant as knocking on a broken jade. The famous doctor Zhou Zhiyuan in the Qing Dynasty wrote in the Chapter of Bamboos and Stones of *Mind of Medical Practitioners: Collected Works*: "By listening to the sound of bamboos, one can clear the mind; by being with bamboos, one can get pleasure. Both ways are better than taking medicine." It shows that the ancients had known the healing function of sound for a long time. At the same time, trees, hedges, and bushes can isolate some noise and provide people with a quiet and relaxing space to help thinking and meditation. With wind chimes or rain bells installed in the garden, the effect of auditory stimulation can be increased. This "zero design" processing method can make people hear real natural sounds and thus achieve relaxation and peace of mind.

除了音乐疗法领域对不同音乐疗愈功能的专业化研究之外，目前学术界在自然声音的健康效益研究方面也积累了大量的循证成果，表明自然声音能够有效降低压力荷尔蒙，减轻支气管镜检疼痛，有助于睡眠和放松、提升情绪、降低疲劳感（Eva，2012）。运用植物媒介有目的地为康复环境营造各种自然声响，是提升环境康复效益的有效手段，已成为疗愈空间设计的基本内容之一。某些医疗机构花园里通常会为孤独症儿童设计听觉疗愈装置，引导其聆听各种隐藏在草丛中的昆虫声音，以达到辅助患儿病症康复疗愈的效果。

In addition to the specialized research in the field of music therapy on the healing functions of different types of music, the current academic community has accumulated a large number of evidence-based results about the health benefits of natural sounds, which show that natural sounds can effectively reduce stress hormones, relieve the pain during bronchoscopy, help sleep and relax, improve mood, reduce fatigue, and so on (Eva, 2012). Using the intermediary effect of plants to purposefully create various natural sounds for the healing environment is an effective means to enhance the environmental healing effect and has become one of the basic contents of the healing space design. In some medical institutions, the auditory healing device was designed in garden for autistic children. The device guides the children to listen to various insect sounds from the grass to achieve the effect of assisting the treatment of the children's illness.

随着现代城市化发展，汽车、飞机、施工、机械设备等噪声已成为人居环境中主要的背景声，给人们带来心血管负担，危及认知和学习效率、降低免疫系统功能，还会导致失眠，诱发抑郁症与焦虑症，是现代城市主要的环境污染之一。如何通过有目的的空间布局，结合针对性的景观设计形成城市绿化空间，不但可为我们提供抵抗噪声的屏障，还能提升环境的声音质量，形成独特的声音疗愈空间。因此，充分了解环境的声音产生及传播机制，充分发挥植物的听觉景观功能，对提升康复空间的疗愈功效具有积极意义。

With the development of modern urbanization, the main background sound in the living environment, noises from automobiles, airplanes, construction, and mechanical equipment bring about cardiovascular burdens, endanger cognition and learning performance, reduce the immune system function, trigger insomnia, and increase the possibility of depression and anxiety. Those noises are one of the main environmental pollution in modern cities. Combining targeted landscape designs to form urban green space through purposeful spatial layout can not only provide us with a barrier against noise, but also improve the sound quality of the environment to form a unique sounds healing space. Therefore, it is of positive significance to fully understand the sound generation and transmission mechanism of the environment and give full play to the auditory landscape function of plants in improving the healing effect of rehabilitation spaces.

3.3.3 嗅觉
3.3.3 Smell

嗅觉是鼻腔里的嗅细胞受到某些挥发性物质刺激而产生神经冲动，这些神经冲动沿嗅神经传入大脑皮质，形成了对各种刺激物的感觉。大量研究成果表明，鼻腔具有强大的"药物"传递功能，是包括荷尔蒙、疫苗、小肽单元混合物等在内的许多物质进入大脑的潜在途径。这些物质经鼻腔进入人体的大脑，渗透全身血液循环，从而对人体发挥作用。花卉所散发的各种袭人香气，可以通过鼻道嗅觉神经直达大脑中枢，能改善大脑功能、激发愉悦感，对疾病康复和预防有一定作用。不同花香对于不同患者有不同的嗅觉刺激效果。另外，花香还可以唤起人们美好的记忆与联想，使人进入冥想状态，以调养身心。

Smell is the sense through which various stimuli are perceived. The olfactory cells in the nasal cavity are stimulated by certain volatile substances to produce nerve impulses, which are transmitted along olfactory nerves to the cerebral cortex. At present, many research results show that the nasal cavity has a powerful "drug" transport function, which is a potential way for many substances including hormones, vaccines, and small peptide unit mixtures to enter the brain. These substances enter the human brain through the nasal cavity and penetrate the blood circulation throughout the body, thereby exerting effects on the body. The various scents emitted by flowers can reach the central nervous system of the brain through the olfactory nerves of the nasal passages to improve brain functions, produce pleasure, having a certain effect on disease recovery and prevention. Different floral scents have different olfactory stimulation effects for different patients. In addition, the fragrance of flowers can evoke people's beautiful memories and associations, making people get into a meditation state to nurse the mind and body.

人类很早就懂得运用芳香植物的提取物。早期的人类将源自树木和花卉的树脂、树胶、油等广泛运用于医疗、宗教、饮食和仪式。古希腊医师知道有些特定植物的芳香能刺激大脑兴奋，而某些种类的植物芳香则具有镇静作用，并根据不同的需要谨慎选择对应的芳香植物进行疗愈。随着科学的进步，科学家对不同植物气味对人体健康的影响机制有了更深入的了解，目前在许多疗愈机构中使用的芳香疗法，就是以现代科学对植物气味及其在人类认知和精神障碍疗愈上的研究成果为依据开展的。例如，几种常用芳香植物及其气味的功效：菊花清热祛风、平肝明目；桂花解郁、避秽，有助于辅助治疗狂躁性精神病；薰衣草安神助眠，有助于缓解头痛、治疗心率过高；天竺葵有助于缓解焦虑及疲劳状态；茉莉花理气、解郁、避秽，可缓解感冒引发的头痛症状。

Humans have long known the application of extracts of aromatic plants. Early humans applied resins, gums, and oils derived from trees and flowers in medicine, religion, diet, and ceremonies. Ancient Greek physicians knew that some specific plant aromas tended to be exciting and stimulating, and that certain types of plant aromas had a calming effect, and they carefully chose corresponding aromatic plants for treatment based on different needs. With the advancement of science, scientists have a deeper understanding of the mechanism of different plant odors' influence on human health. The aromatherapy currently used in many healing institutions is based on modern science's research results on plant odors and their role in healing cognitive and mental disorders. Several commonly used aromatic plants and their odor effects are as follows: The Chrysanthemum has the effect of clearing away heat, expelling wind, calming the liver and improving eyesight; the Sweet-scented Osmanthus relieving depression, avoiding filth, and helping to assist in the treatment of manic psychosis; the Lavender soothing the nerves, helping sleep and relieve headaches, treating tachycardia; the Geranium helping relieve anxiety and fatigue; the Jasmine regulating the flow of qi (vital energy), relieving depression, avoiding filth, and relieving headache symptoms caused by a cold.

植物释放的芳香物质中蕴含的植物杀菌素能够降低应激激素的产生，减轻焦虑，增大疼痛阈值，同时增强人体抗氧化防御系统。此外，植物杀菌素还与人体的免疫功能有密切关联。日本李卿教授团队对森林环境的健康效应进行了系统研究，发现森林中大量传播的植物杀菌素会刺激血液中产生更多的抗癌蛋白(李卿，2013)，激发人体更高水平的前锋免疫防御，即自然杀伤细胞(NK 细胞)。当人们暴露于病毒(如流感、感冒)和其他传染物时，自然杀伤细胞就会行使保护功能，因此自然杀伤细胞活性越高的成年人，感冒等疾病的患病频率就越低。还有研究显示，抑郁症患者嗅球(负责嗅觉)更小，察觉空气中微弱气味的能力下降，因此其感知到的世界与常人不同。有目的地配置各种芳香植物，为抑郁症患者提供适当的嗅觉刺激，对缓解抑郁情绪是有益的。也有研究将嗅觉与其他感官功能障碍的辅助疗愈结合起来，或者研究嗅觉对改善焦虑症及注意力不集中症的积极效益等。这些研究成果进一步阐释了植物气味对人体健康的积极效益，为现代康复环境的植物气味设计提供了循证依据。

Plants can release very rich aromatic chemicals. The phytoncide contained in those substances can reduce stress hormones and anxiety and increase the pain threshold, while enhancing body's antioxidant defense system; in addition, the phytoncide is also closely related to physical immune function. The team of professor Li Qing from Japan conducted a systematic study on the healing effects of the forest environment and found that many phytoncides spreading in the forest cause the production of more anti-cancer proteins in the blood (Li Qing, 2013) and stimulate a higher level of frontline immune defenders, that is, natural killer cell(NK cell). When people are exposed to viruses (such as flu, colds) and other infectious agents, natural killer cells will come out to protect us. Therefore, the more active the Natural Killer (NK) cells are in adults, the less frequently they tend to suffer from illnesses such as the common cold. Other studies have shown that the olfactory bulb (responsible for smell) of patients with depression is smaller and their ability to detect faint odors in the air is reduced, so the world they perceive is different from what ordinary people can. The purposeful arrangement of various aromatic plants to provide appropriate olfactory stimuli for patients with depression helps alleviate depression. There are also studies that combine the auxiliary treatment of the smell dysfunction and other sensory dysfunctions, or study the positive effects of the sense of smell on improving the anxiety disorder and the attention deficit hyperactivity disorder. These research results have further clarified the benefits of plant odors for human health, and provided the evidence-based basis for the design of plant odors in modern healing environments.

必须注意的是，也有部分植物所散发的气味对人体有害，或者引发某一类患者身体不适。一些特殊患者，如刚刚经过化疗的癌症患者对气味十分敏感，芳香型植物不适宜种植在针对这类患者的康复花园中。

It is noteworthy that the odors emitted by some plants are harmful to the human body or cause discomfort to certain types of patients. Some special patients, like the patients with cancer who have just undergone the chemotherapy, are very sensitive to odors, and it is not suitable for aromatic plants to be planted in the healing gardens that serve this type of patients.

3.3.4　味觉
3.3.4　Taste

味觉是由食物对口腔内味觉器官的化学感受系统进行刺激所形成的感觉，往往与嗅觉共同作用。某些疾病，或者随年龄增长导致的身体机能衰退，都会引发紊乱的神经元冲动散播、神经递质功能改变，以及口腔黏膜干燥等，使得与味觉相关的高级感觉信息过程改变，从而导致味觉障碍。因此，味觉感受在一定程度上可以反映人体的健康状况，其相关功能的保护与康复对保障人体健康至关重要。

The sense of taste is formed by food's stimulating the chemosensory system of taste organs in the oral cavity, and it often acts together with the sense of smell. Certain diseases, or the decline of body functions resulting from aging can cause the dissemination of disordered neuronal impulses, changes in the neurotransmitter function, the dry oral mucous membranes and so on, thereby changing the high-level sensory information process related to taste and leading to taste disorders. For this reason, the sense of taste can reflect the health status of the human body to a certain extent, and the protection and rehabilitation of its related functions are essential to safeguarding human health.

植物是提供人类食物的主要途径，种植一些可食用的植物，诸如果树、蔬菜、农作物等，让患者亲身体验从种子到餐桌的食物成长过程，不但可以加深患者对食物的味觉体验，还能通过品尝自己的劳动成果，产生极大的满足感和成就感，增强战胜疾病的信心；培育一些具有药用价值的植物，尤其是对患者疾病有治疗价值的植物，可以帮助患者了解疾病治疗的药物机制，从中获得战胜疾病的力量，对患者情绪提升有积极影响。

Plants are the main way to provide human food. By planting some edible plants, such as trees, vegetables, grains and so on, patients can experience the growth of food from seed to food served at the table, which not only enrich the patient's food tasting experience, but also make the patient generate a great sense of satisfaction and accomplishment and enhance the confidence in overcoming disease by tasting the fruits of his/her own labors. Cultivating some herbal medicines with medicinal value, especially herbal medicines with therapeutic value for patients' diseases, can help patients understand the medicines' disease treatment mechanism, from which the power to overcome the disease is obtained, and which has a positive effect on the emotional improvement of patients.

食物从种子到餐桌需要一个很长的生产过程，这就为参与者提供了一个围绕共同目标合作互助的机会。人们在康复花园中栽种水果、蔬菜、香草，食用成熟的鲜果；收割成熟的瓜菜，可一起烹调和享用；采摘食用香草，用作食材或冲泡花茶，均能产生味觉治疗效果。也可以栽植一些食用花卉，既可以用作食材，又可以全花入药，或提取花粉和花蜜。总之，味觉刺激可以达到有效提升参与者的成就感和满足感的目的。在社区种植可食地景，可以促进公众参与，增强邻里之间的情感联系，有利于形成健康和谐的社区关系，对促进人体健康具有十分重要的意义。第10章将对可食地景的营建做专门阐述。

From seed to table, food needs a long production process, and for this reason, participants are provided with an opportunity to cooperate and work for a common goal. People plant fruits, vegetables, herbs and eat ripe fruit in the healing garden, harvest ripe melons and vegetables to cook together and enjoy them, pick edible herbs to add them to recipes or brew them to make scented tea to produce a taste healing effect. Some edible flowers can be planted to be used as food; their whole flowers can be used as medicine or for extracting pollen and nectar. In short, taste stimulation can achieve the purpose of effectively enhancing the participants' sense of accomplishment and satisfaction. With edible landscapes created in the community, the public participation can be promoted and the emotional connection among neighbors strengthened, which is conducive to the formation of healthy and harmonious community relations and is of great significance to the promotion of human health. In the tenth chapter, we will elaborate on the construction of edible landscapes.

3.3.5 触觉
3.3.5 Touch

触觉是与人类情感联系最密切的感觉系统，是接触、滑动、压迫等机械刺激的总称。人体主要的触觉器官位于体表的皮肤，通过表皮游离的神经末梢，可感受温度、疼痛、振动等遍布全身的多种感觉。研究表明，婴儿时期较多得到亲人抚摸的孩子，成人之后人格更加健全，智力发育也更完善。

The sense of touch is the most closely related to human emotions and is the general term for mechanical stimuli such as contact, sliding, and pressure. The main tactile organ of the human body is the skin on the surface of the body. Through the free nerve endings of the epidermis, various sensations throughout the body such as temperature, pain, and vibration can be felt. Studies have shown that children who are touched more by their relatives during infancy have a more wholesome personality and better intellectual development in adulthood.

植物种类繁多，其枝干、表皮、叶片，乃至花朵、果实、种子等均具有独特的生命特征，通过触摸不但可以获得关于植物形状、质感、表面温度、湿度等特征的不同感官刺激体验，如平滑、粗糙、茸毛、坚实、薄脆、肉质等，还有助于提高参与者的认知能力，加深对自然生命的直观理解。这对促进儿童认知能力发育，提升其思维水平与分析能力等意义重大，是儿童自然教育中不可或缺的内容之一；老年人随着年龄增长，身体机能发生退化，或因疾病引发各种机能衰退，通过抚摸植物，不但可以增进触觉功能，还能体验生机勃勃的生命乐趣，激发身心内在的各种积极反应，对患者生命质量的提升是全方位的，远非其他作业疗法所能企及。

There are many kinds of plants, and their branches, barks, leaves, and even flowers, fruits, seeds, and all have unique life characteristics. You can get different sensory stimulation experiences (like smoothness, roughness, fluff, firmness, crispness, flesh, and so on.) about plants' shape, texture, surface temperature, humidity and other characteristics by touching them, which helps to improve the cognitive ability of the participants and deepen the intuitive understanding of the life in nature. This is of great significance for promoting the development of children's cognitive ability and improving their thinking level and analytical ability. It is part of the indispensable contents in children's natural education. As the elderly grow older, their physical functions would degenerate, or diseases may cause their hypo function. But by touching plants, the function of the tactile sense can be improved, and patients can experience the vitality of life with various positive reactions stimulated within the mind and body, and improve their quality of life in an all-round way. Such a method is far better than other occupational therapies.

由于触摸是一种与肉体直接接触的行为，用于触摸的疗愈植物不但要求质感丰富、生命力强，还必须保证是无毒、无刺的种类。由于适宜不同地域种植的植物有一定差异，充分了解所处地域的植物及其基本特征，从中选择适宜的康复疗愈植物至关重要。

Since touching is an act of direct contact with the body, the healing plants used for touching must have rich texture and strong vitality and be non-toxic and thorn-free. As there are certain differences in the plants suitable for being planted in different regions, it is vital to fully know the plants and their basic characteristics in the region and choose suitable healing plants from them.

3.4　循证设计的基础
3.4　The basis of evidence-based designs

3.4.1　循证设计的概念
3.4.1　The concept of evidence-based designs

循证设计中"循证"的概念最早起源于医学领域中的"循证医学"（EBM）理论，循证医学是对"临床证据"进行辨析和归纳的医学。循证设计的概念在 2003 年由医院建筑师、学者 Kirk Hamilton 提出，他认为循证设计是在实践项目的重要设计决策中，严谨、认真地去找寻现存的相关研究和实践产生的最佳设计证据（Stichler，2008）（图 3-4）。美国加利福尼亚州的"健康设计中心"（CHD）将这一定义进行叙述为：将环境的设计决策建立在可信的研究之上，从而达到最佳预期结果的一种过程。后来我国的建筑学者将循证设计的概念引入国内，并在医疗建筑设计中加以实践应用（刘玉龙，2006）。我国风景园林的相关学者也利用循证设计的研究思路，将其拓展到康复景观的设计中，用来探寻促进不同人群的康复设计线索与证据（郭庭鸿，2015；贾君兰，2020）。表 3-2 对比了循证设计与传统设计的部分差异。

The concept of "evidence-based" in evidence-based designs originated from the theory of "evidence-based medicine" (EBM) in the medical field. The evidence-based medicine is the medicine that differentiates, analyses and sums up "clinical evidence". The concept of the evidence-based design was proposed by hospital architect and scholar Kirk Hamilton in 2003. He believes that the evidence-based design, among important design decisions for practical projects, is to rigorously and seriously look for the best design evidence generated by existing related researches and practice (Stichler, 2008) (See Fig. 3-4). The Center for Health Design (CHD) in California, USA described this definition as: a process of basing environmental design decisions on credible researches to achieve the best expected results. Later, Chinese architecture scholars introduced the concept of the evidence-based design to China and applied it in medical building designs (Liu Yulong, 2006). Landscape architecture scholars in China used the research ideas of the evidence-based design and extended them to the design of healing landscapes to explore the clues and evidence of healing designs for helping different groups of people (Guo Hongting, 2015; Jia Junlan, 2020). Table 3-2 presents a comparison of selected differences between traditional design and evidence-based design.

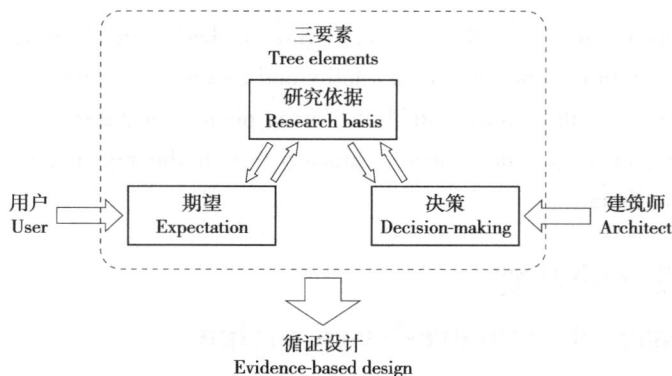

图 3-4　循证设计三要素

Fig. 3-4　Three elements of evidence-based designs

（图片来源：王一舒改绘自《循证设计：含义解析》）

(Image source: from *Evidence-based Design: What Is It*? redrawn by Wang Yishu)

表 3-2 传统设计和循证设计

Tab. 3-2 The traditional design and evidence-based design

评价指标 Evaluation Criteria	传统设计 Traditional design	循证设计 Evidence-based design
证据来源 Source of evidence	场地调研、资料查找、教科书、零散案例研究 Site survey, data search, textbooks, scattered case studies	强调科学研究 Emphasize scientific research
收集证据 Collect evidence	不够系统全面 Not systematic and comprehensive	系统全面 Systematic and comprehensive
评价证据 Evaluate evidence	不够重视 Not value evidence	重视 Value evidence
判断依据 Judgment basis	甲方满意度 Party A's satisfaction	场地使用者体验 Experience of users on site
设计模式 Design mode	设计师/甲方为核心 Designer/Party A is the core	场地使用者为核心 Users on site are the core

3.4.2 循证设计的研究框架
3.4.2 The research framework of evidence-based designs

循证设计中核心要素是找寻设计证据，证据的生产、评价、传播、使用与反馈构成了循证设计实践的整个过程。处于起步阶段的循证设计常常采用定性和定量的方法来研究（Viets，2009）。完整的循证设计是一个循环反馈的过程，包括提出假设、试证研究假设、设计实践以及设计建成之后的适用性评估（图 3-5）。

The core element of evidence-based designs is to find design evidence. Production, evaluation, dissemination, use and feedback of evidence constitute the entire process of the evidence-based design practice. The evidence-based design in its initial stage often uses qualitative and quantitative methods to carry out a research (Viets, 2009). A complete evidence-based design is a cyclic feedback process, including proposing a hypothesis, validating a research hypothesis, the design of practice, and the applicability evaluation after the design is completed (Fig. 3-5).

3.4.3 循证设计的研究方法
3.4.3 Research methods of the evidence-based design

循证设计的研究方法主要是根据循证数据来源形式进行划分，主要分为定性和定量两种。

The research methods of evidence-based designs are mainly classified into qualitative methods and quantitative ones according to the form of evidence-based data sources.

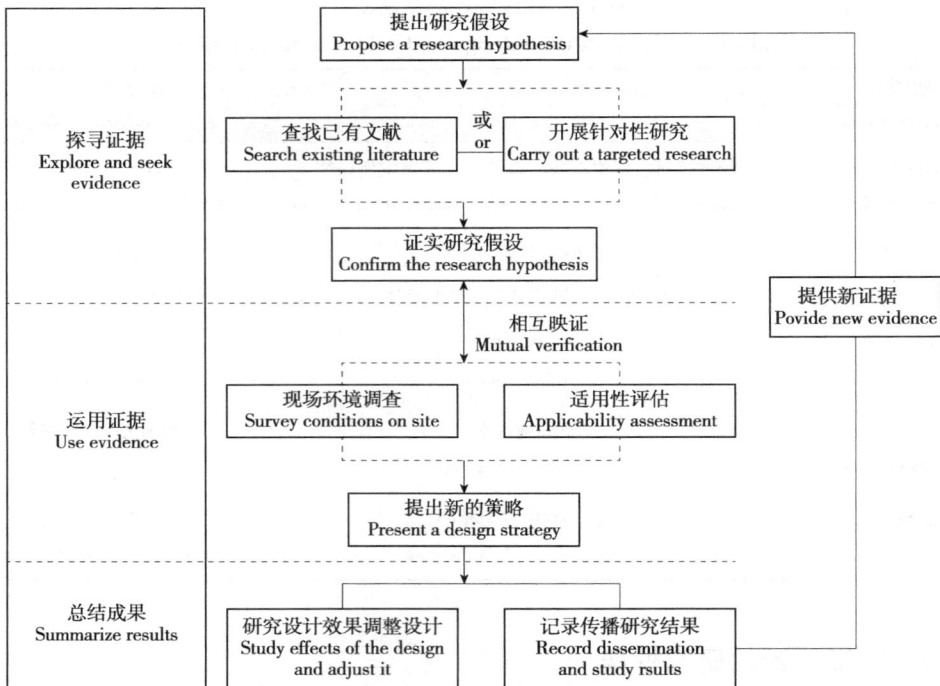

图 3-5 循证设计研究框架

Fig. 3-5 The research framework for evidence-based designs

（图片来源：王一舒改绘自《设计与实证：康复景观的循证设计方法探析》）

（Image source：from *Design and Evidence*：*Analysis on the Method of Evidence-based*

Therapeutic Landscape Design，redrawn by Wang Yishu）

3.4.3.1 定性研究法

3.4.3.1 The qualitative research methods

定性数据收集，通过对设计使用主体的行为进行观察和访谈，或者查阅相关参考文献和视听材料，以获取有关设计要素合理化的有效信息。例如 Masashi Soga 等就在《园艺有益于健康：荟萃分析》一文中研究了园艺对公众健康影响的证据（Soga，2017），该研究首先搜索收集了大量文献，比较了 2016 年 1 月对照组（参加园艺活动或非园艺工作者之前）和治疗组（参加园艺活动或园艺工作之后）的健康结果。每项研究计算两组间健康结果的平均差异，然后确定所有子组研究和各组研究的加权效应大小。研究报告了一系列广泛的健康结果，如抑郁、焦虑和体重指数的降低，以及生活满意度、生活质量和社区意识的提高，并分析园艺活动对于公众健康的影响。

Qualitative data collection means obtaining effective information about the rationalization of design elements by observing the behavior of the design users and interviewing them，or by consulting some related reference documents and audio-visual materials. For example，Masashi Soga studies the evidence of the impact of gardening on public health in her paper *Gardening is beneficial for health*：*A meta-analysis* (Soga，2017). In the study，a literature search was performed first to collect studies that compared health outcomes in control（before participating in gardening or non-gardeners）and treatment groups（after participating in gardening）in January, 2016. Each study calculates the average difference in health outcomes between the two groups, and then determines the weighted effect size for all subgroup studies and each group study. Studies have shown a wide range of health outcomes, such as reductions in depression, anxiety, and body mass index, as well as increases in life satisfaction, quality of life, and sense of community and analyzed the impact of gardening activities on public health.

3.4.3.2　定量研究法
3.4.3.2　The quantitative research methods

定量的研究包括公众针对设计场地诉求的问卷调查反馈，或者通过随机对照试验（RTC）（"假说—演绎"），通过量表、身心反馈仪等实验仪器来研究设计是否满足使用者的需求。基于此，园艺康健成了康复花园循证设计定量研究的主要手段，通过园艺康健的实施，方可从认知科学的角度去了解环境因素对人类行为、情感和生理的影响。

Quantitative research includes questionnaire feedback of the public on the demands of the design site, or randomized controlled trial（RTC）（"hypothesis-deduction"）and experimental instruments such as scales and physical-mental feedback devices to find out whether the design can satisfy the users' needs or not. Based on this, the horticulture therapy has become the main means of quantitative research in the evidence-based design of healing gardens. Only through the implementation of the horticulture therapy can we understand the impact of environmental factors on human behavior, emotion and physiology from the perspective of cognitive science.

思考题

1. 如何理解中国人养花赏花修身养性的传统文化与现代康复花园设计之间的关系？
2. 请阐述指导康复花园设计的主要理论。
3. 为什么康复花园设计要强调循证研究？

Questions

1. How do you understand the relationship between the Chinese traditional culture of cultivating and admiring flowers for self-cultivation and the design of modern healing gardens?

2. Please explain the main theories guiding the design of healing gardens.

3. Why should the design of healing gardens emphasize evidence-based researches?

第4章

园艺康健概述

Chapter 4　Overview of Horticultural Therapy

关于园艺康健，美国园艺康健协会将其定义如下：对于有必要在身体以及精神方面进行改善的人们，在园艺康健师带领下，通过实施对症的园艺操作活动或者与植物相关联的各种活动，以达到使参与人恢复身体机能、恢复心理健康以及恢复对社会的适应力等目标的方法（美国园艺康健协会官方网站，2017）。园艺康健也可以理解为人们通过各种园艺活动，达到促进身心健康、消除疲劳、增进思维、益寿延年的目的。康复花园与园艺康健是一对不宜分割的概念，没有园艺康健的康复花园是缺乏内涵的，难以使人达到真正康复的有效目标；而没有康复花园的园艺康健则少了环境的依托，略显单调。因此，在人通过照护植物的过程去实现自身身心健康的这个模式中，康复花园与园艺康健成了一对不可或缺、互为支撑的硬件与软件，即园艺康健采用科学的活动安排为康复花园的康复性功能提供软件支持。

According to the American Horticultural Therapy Association（AHTA）, horticultural therapy is defined as a therapeutic modality for participants who need physical and mental improvement, and which helps participants achieve the goals such as recovering physical and mental health and restoring the adaptability to the society, under the guidance of horticultural therapist in horticulture practices or plant-related activities（The official website of the American Horticultural Therapy Association, 2017）. It can also be viewed as a process aimed to improve physical and mental wellbeing, relieve fatigue, sharpen mind and prolong life by means of horticulture practice. Healing gardens and horticultural therapy（HT）are not to be discussed separately. A healing garden without therapeutic horticultural activities wouldn't live up to its name, and it is hard to achieve real rehabilitative goals. Meanwhile, the horticultural therapy without healing gardens as the environmental support would be a bit monotonous. Therefore, in the model of improving people's physical and mental health by tending plants, healing gardens and horticultural therapy serve as the hardware and software that are indispensable and mutually supportive. In other words, horticultural therapy, with scientifically-designed activities, can provide software support for the rehabilitative functions of healing gardens.

4.1 园艺康健概念及起源
4.1 Definition and origin of horticultural therapy

园艺康健是通过训练有素的园艺康健师引领参与者直接或间接与植物进行互动，以达到恢复身体机能、恢复心理健康以及增进社会适应能力等目标或达到某个明确特定的治疗目的的方法。美国园艺康健协会认为园艺康健是一个建立在完整治疗方案之上的园艺活动过程(American Horticultural Therapy Association, 2018)。

Horticultural therapy refers to the treatment modality in which well-trained horticultural therapists guide participants to interact with plants directly or indirectly, in order to rehabilitate physical functions, restore mental health and improve social adaptability, or to achieve a specific therapeutic purpose. The American Horticultural Therapy Association (AHTA) defines horticultural therapy as the participation in horticultural activities within an established treatment plan (AHTH, 2018).

早在古代中东地区，人们就建造色彩鲜艳、花香馥郁的园林用以愉悦身心、调节情绪(李树华, 2016)。现代园艺康健的概念起源于 17 世纪末的英国，但其作为一个现代行业得以发展壮大却是在美国。1798 年，美国医生本杰明·拉什因发现在庭院工作的一名精神患者病状减轻，首次报道园艺活动可以缓解精神不安与神经系统症状(郑丽, 2016)。此后在欧美诸国，园艺康健时常用于疗愈精神疾病。在第一次世界大战和第二次世界大战中，伤者常通过在医院等机构进行园艺活动，进行肢体的康复训练和心理创伤治愈，与此同时兼顾农业生产。第二次世界大战后，园艺康健作为退役军人康复治疗和社会再适应的重要手段之一被广泛采用。1973 年，堪萨斯州立大学首次开设了心理学和园艺学交叉的课程体系(Simson, 2003)。20 世纪 80 年代以后，随着压力缓解理论(SRT)和注意力恢复理论(ART)的出现和普及，园艺康健的作用机理和具体效果逐渐明了。目前，世界上很多国家设有园艺康健学会或协会，以推进园艺与健康的研究和实践。

In the ancient Middle East, people built gardens with colorful and fragrant flowers to please their bodies and minds and regulate their moods (Li Shuhua, 2016). The concept of modern horticultural therapy originated in England in the late 17th century, but it grew and thrived as a modern industry in the United States. In 1798, an American doctor, Benjamin

Rush, reported for the first time that gardening activities could relieve mental disturbance and neurological symptoms after he found a psychopath working in a garden showed improved symptoms (Zheng Li, 2016). Since then, horticultural therapy has been frequently used to treat mental illness in Europe and the United States. During World War I and World War II, the injured often received physical rehabilitative training and psychological trauma healing in hospitals and other institutions by participating in gardening activities, and agricultural production. Over the postwar decades, horticultural therapy was widely adopted as one of the important means of rehabilitation and social adaptation for veterans. In 1973, Kansas State University offered the first cross-curriculum of psychology and horticulture (Simson, 2003). After the 1980s, with the emergence and wide recognition of Stress Reduction Theory (SRT) and Attention Restorative Theory (ART), the mechanism and specific effects of horticultural therapy gradually became clearer. At present, horticultural therapy societies or associations have been established in many countries to promote the research and practice of horticulture and health.

4.2 园艺康健作用途径及优势
4.2 Functioning and advantages of horticultural therapy

园艺活动是一种人与植物直接或间接发生互动的复杂过程。在这一过程中，活动参与者的心理、生理、人际关系、身处环境等方面均受到影响并发生改变。这一系列变化共同作用，最终表现为园艺康健疗效(图4-1)。

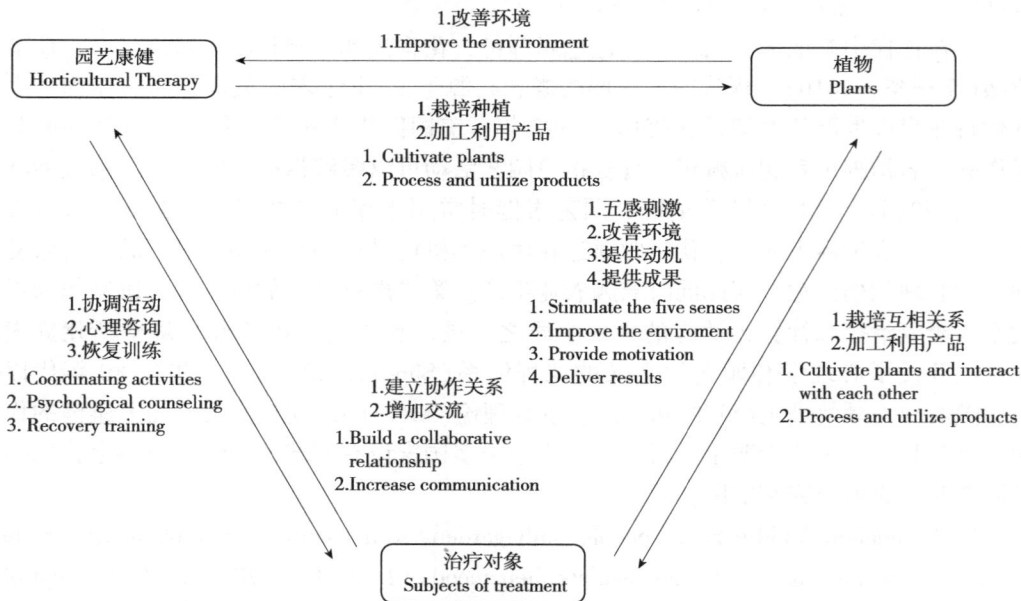

图 4-1 园艺康健示意图

Fig. 4-1 Schematic diagram of curative effect of horticultural therapy

Horticulture is a complex process in which people interact with plants directly or indirectly. In this process, the psychological and physiological conditions, interpersonal relationships and environments of the program participants are influenced and altered. The interaction of these changes contributes to the ultimate efficacy of horticultural therapy (Fig. 4-1).

4.2.1 园艺康健的作用途径
4.2.1 Functioning of horticultural therapy

园艺康健是一系列与植物相关活动的综合体，其对人身心健康的影响途径主要有以下三个方面。

Horticultural therapy is a combination of a series of plant-related activities, and its effects on human physical and mental health mainly lie in the following three aspects.

（1）身体活动

（1）Physical activity

栽种、管理、收获等，都要求服务对象进行一定程度的身体活动以完成指定流程。这一系列的身体活动强度中等或较低，可锻炼的身体部位覆盖面广，与日常生活身体活动所需能力 ADL 相似，适宜作为身体能力恢复训练的一环。同时，适度的身体活动可以帮助服务对象调节自主神经系统和内分泌系统，对心理健康和免疫力的维持或恢复有积极效应。

Planting, management and harvest engage the program participants in certain physical activities to complete the specified process. These activities are of medium or low intensity while exercising a wide range of body parts, similar to the activities of daily living (ADL). Therefore, they are appropriate as a part of physical ability recovery training. At the same time, moderate physical activity can help patients regulate their autonomic nervous system and endocrine system, exerting a positive effect on the maintenance or recovery of mental health and immunity.

（2）感官刺激

（2）Sensory stimulation

与植物互动是园艺康健的核心内容，植物与生俱来的色彩、触感、气味、味道以及与周围环境发生接触时发出的声音会刺激和滋养服务对象感官。园艺活动一般在室外或者半室外空间进行，与室内空间相比，实施园艺康健的场所即康复花园，通常为花草树木所环绕，生机勃勃。植物的主色调绿色，可使人主观感到自然、放松；花卉果实等各具鲜艳色彩，可以提供良性视觉滋养。

Interaction with plants is the core of horticultural therapy. The plants' colors, touch, smell, taste and the sounds they make in contact with the surroundings will stimulate and nourish the feelings of patients. Horticultural activities are generally carried out in outdoor or semi-outdoor spaces. Unlike indoor sites, a horticultural therapy garden is usually surrounded by flowers and trees and full of vitality. The green color, the major tone of plants, makes people comfortable and relaxed. Bright-colored flowers and fruits also provide positive visual nourishment.

花卉尤其是香草植物散发出自然芳香，对于缓解压力、减轻疲劳、调节自主神经活动均有一定功效。加工食用蔬菜、水果等刺激对服务对象的味觉，味道佳的植物可提升服务对象的成就感，提高服务对象继续治疗的积极性；若味道不佳，究其原因多数是服务对象的食用方式有误或食用了非食用植物，这一类负面刺激可以帮助服务对象重新回想起生活常识，增强记忆回溯能力，提高日常生活能力。

The natural fragrance of flowers, especially aromatic plants, helps relieve stress, reduce fatigue and regulate the autonomic nervous activities. Processed vegetables and fruits can stimulate the participants' gustation. A good flavor can provide a sense of achievement and duly motivate the patients to continue treatment. An unpleasant taste, usually due to improper ways of processing or eating inedible plants, can be a negative stimulus that helps the patients recall common sense in life, enhance memory retrospection, and improve the abilities needed for daily living.

进行园艺活动时，手脚与植物材料、工具以及周边环境如土壤等发生接触所产生的各类不同触感，对服务对象是一种良性的感官刺激。此外，身处室外空间进行园艺活动时，如微风拂过植物枝叶时发出的沙沙声、浇水时的水声以及雨天进行园艺活动时的雨声等源于自然的声音造成的听觉刺激可以调节自主神经活动、稳定情绪。

In gardening, the different tactile senses, produced by the contact of hands and feet with plant materials, tools and the surroundings such as soil, are positive sensory stimuli to the participants. In addition, when the participants conduct gardening activities outdoors, auditory stimuli from the natural sounds, like rustles of breeze blowing through branches and leaves, gurgles of watering the plants and drops of rain pitter-pattering on a rainy day, can regulate their autonomic nervous activity and soothe their emotions.

（3）社交

（3）Socializing

园艺康健实施过程中，园艺康健师作为合作伙伴，一起参与园艺活动。在此过程中，植物是园艺康健师和服务对象共同的主要话题，可以引导双方顺畅地交流和相互理解。对于存在心理健康问题的服务对象，以植物为介质可以引导其更好地接受心理治疗。园艺康健流程往往会包含多个服务对象，不同对象可以有不同的治疗方式，根据园艺康健流程设计，植物同样能为服务对象之间的交流提供话题。

During a horticultural therapy program, the therapist, as a partner, works with the patient in horticultural activities. In this process, plants are the shared topic for the therapists and patients, leading them to communicate smoothly and understand each other better. For participants with mental problems, using plants as the medium helps them to enter psychotherapy more easily. A horticultural therapy often involves a group of participants.

There can be different therapies for different participants. According to process design for horticultural therapy, plants can serve as topics for communication among the participants.

4.2.2 园艺康健的优势
4.2.2 Advantages of horticultural therapy

首先，园艺康健可以提高服务对象参加治疗的积极性，使其配合治疗。对于服务对象而言，进行康复治疗，促进身心健康需要较强的主观能动性，其他疗法如物理疗法等往往伴随着枯燥的治疗过程，服务对象缺乏足够的主动参与性。与之相比，园艺康健的治疗过程即植物栽培和利用的过程，同时包含较为休闲的目的性以及对生产结果的期待感，使得服务对象更易于主动配合参与治疗，提高治疗效果。因此，园艺康健的优势主要表现在为服务对象提供充分的目的性和积极性。服务对象自主地参加活动，通过与植物互动得到喜悦感和成就感，互动过程中产生的身体活动，与植物接触时受到的感官刺激以及活动中发生的社交行为是服务对象恢复身心健康的关键。

First of all, horticultural therapy can motivate the patients to participate in treatment and cooperate with the therapist. Rehabilitative treatment to improve physical and mental health requires a comparatively strong initiative of the program participants. Other treatment modalities, like physiotherapy, are often accompanied by boring treatment process, greatly discouraging the participants. In contrast, horticultural therapy is the process of plant cultivation and utilization, with recreational purposes and the expectation for the end products, making the participants more cooperative in treatment and thus producing better therapeutic effects. Therefore, the biggest merit of horticultural therapy is the purposefulness and enthusiasm it prompts among the participants. They take part in the activities voluntarily. The key to restoring the participants' physical and mental health lies in the joys and a sense of accomplishment, physical activities, sensory stimulation and socialization in their interaction with plants.

其次，基于SRT理论、ART理论以及诸多基础研究结果，人与自然环境有着密不可分的亲密关系，人对植物的亲近感与生俱来。因此园艺康健与其他疗法不同，受患者个人好恶的影响程度较小，园艺康健的服务对象不易因产生厌倦感而使得治疗效果不佳。

Secondly, according to SRT theory, ART theory and the findings of many basic researches, there is a close relationship between people and the natural environment, and people are born with an intimacy with plants. Compared with other treatment modalities, horticultural therapy is affected less by patients' personal preferences, so they are less likely to experience boredom, which may impair the healing effects.

园艺康健实施过程中，园艺康健师与服务对象之间存在着植物这一介质，因此整个过程中园艺康健师与服务对象的关系并非传统康复治疗中的给予与接受的关系，而是共同朝着一个目标(植物的产出)努力的伙伴关系，因此治疗师和服务对象之间更容易建立信赖关系。同时，在共同参与园艺康健的多名服务对象之间，也更容易产生正向交流，帮助某些有心理问题的服务对象缓解孤独，恢复社会适应能力。

In the process of horticultural therapy, plants serve as a medium between the therapist and the program participants, so their relationship is not giving-and-receiving as in traditional rehabilitative therapies, but a partnership moving towards a common goal (plant production). Therefore, it is easier for them to establish trust. At the same time, HT program participants are more likely to have positive communication, which helps the patients with mental problems alleviate loneliness and restore social adaptability.

园艺康健过程中园艺活动的目标明确、活动成果容易理解，服务对象对园艺活动的最终产品及使用方式相对明确。因此园艺康健较其他疗法而言，更能激发服务对象对未来的期待感和生存欲望。同时，园艺产品尤其是农产品，大多可以直接使用在日常生活中，可以帮助服务对象增强自信、重新认识自己的社会贡献和价值。表4-1列出了园艺康健与其他作业疗法的对比。

With definite goals and conceivable outcomes, the participants are clear about the final products of horticultural activities and the ways of using them. Therefore, compared with other treatments, horticultural therapy can better stimulate the patients' expectation for the future and desire for survival. At the same time, most horticultural products, particularly agricultural produce, can be directly used in daily life, which helps patients improve their confidence and re-recognize their social contribution and value. Tab. 4-1 shows the comparison between horticultural therapy and other occupational therapies.

表 4-1　园艺康健与其他作业疗法的对比

Tab. 4-1　Comparison between horticultural therapy and other occupational therapies

评价维度 Evaluation dimensions	园艺康健 Horticultural therapy	其他作业疗法 Other occupational therapies	园艺康健的优势 Advantages of horticultural therapy
目的性 Purposefulness	目的明白易懂、产出明了，用途明确 HT has clear purposes, conceivable output and definite uses	有些作业疗法产出不明确，产出用途有限 The output of some occupational therapies is unclear, and the uses of the output are limited	园艺康健的产出多与日常生活直接相关，易于激发服务对象的积极性 The output of HT is closely related to daily life, and it is easy to stimulate the patients' enthusiasm
普适性 Applicability	人与植物有天生的亲近感，受个人好恶影响较小 People are born with an intimacy with plants, so HT is affected less by patients' personal preferences	疗效受个人喜好左右 The efficacy is influenced by personal preferences	园艺康健接受度高，方便大规模推广 HT is more acceptable and easier to apply on a large scale

（续）

评价维度 Evaluation dimensions	园艺康健 Horticultural therapy	其他作业疗法 Other occupational therapies	园艺康健的优势 Advantages of horticultural therapy
难易度 Difficulty	多数适用植物种植难度小，参与门槛低 Most suitable plants are not difficult to grow, with a low threshold for participation	有些作业疗法需要一定的基础技能 Some occupational therapies require certain basic skills	园艺康健实施简单，便于开展 HT programs are easy and convenient to carry out
心理层面 Psychology	与植物互动能够使人内心平稳、减轻压力、恢复注意力等 Interaction with plants can calm the mind, reduce stress, and restore concentration	某些作业疗法容易引起服务对象紧张、焦虑、无力感 Some occupational therapies tend to cause tension, anxiety and inferiority of the patients	园艺康健对促进身心健康具有较好的效果 HT is effective in improving physical and mental health
生理层面 Physiology	管理植物所需身体活动基本包含了日常生活所需 The physical activities involved in plant management cover almost all the needs of daily life	某些作业疗法的身体活动未必与日常生活直接相关 The physical activities of some occupational therapies may not be directly related to daily life	日常生活所需的身体活动能力可直接受益于园艺康健所含的身体锻炼内容 The physical activities involved in HT programs can directly benefit the physical performance required in daily life

4.3 园艺康健主要效果
4.3 Major healing effects of horticultural therapy

园艺康健的作用主要包括两条途径：①园艺活动，包含栽种、管理、收获、使用、加工植物；②接触自然，包括接触、观赏自然环境和园林景观。通过这两条途径，园艺康健对服务对象的认知能力、生理健康、心理健康、社会适应性带来显著正面影响。园艺康健的上述作用随着服务对象的不同群体状况（老人、儿童、学生、高压人群等）带来不同影响。园艺康健的效果涵盖多个方面，相互融合。基于循证理念，将园艺康健效果分为以下四个方面进行阐述。

The efficacy of horticultural therapy is achieved mainly in two ways：①horticultural activities, including growing, management, harvest, using and processing of plants；②contact with nature, including contact with and appreciation of the natural environment and landscape. Through these two ways, horticultural therapy has a significantly positive effect on the patient's cognitive ability, physical health, mental health and social adaptability. The above-mentioned effects vary with different patient populations (the elderly, children, students and those under high pressure). The effectiveness of horticultural therapy covers several aspects that integrate with each other. By the evidence-based approach, they fall into the following four categories.

4.3.1 改善认知能力
4.3.1 Improving cognitive functioning

卡普兰等指出，暴露在自然环境中可以在一定程度上增强人的认知能力（Kaplan，1989）；身处自然环境之中比处于城市环境之中能帮助人更好地恢复注意力（Herzog，1997）。在自然环境中实施园艺康健可以帮助服务对象增强集中力、提高注意力水平（Irvine，2002；Laumann，2003）。研究显示，与短期园艺康健相比，实施中长期园艺康健可以显著提高失智老人的认知能力（杉原式穗，2011）。实施中期、频度中等的园艺康健方案（如 4 个月、每月 2 次、每次 1 小时），在疗程结束后，服务对象（阿尔茨海默病患者）的简易智力状态检查量表（MMSE）得分增加（寺冈佐和，2003）；实施长期、高频度的园艺康健（如 6 个月、每月 3~4 次、每次 30 分钟）后，服务对象（阿尔茨海默病患者）的 MMSE 得分与疗程开始时保持同等水平，未进一步下降，并且在园艺康健疗程结束后，MMSE 得分维持同等水平 3 个月以上，重度阿尔茨海默病患者在接受园艺康健治疗后，记忆力、记忆回溯能力有明显提升（孙旻恺，2018）。更多研究显示，园艺康健治疗可以有效帮助儿童提高注意力和记忆力（Taylor，2001；Kima，2012；代星，2018）。另外，园艺康健、植物所散发的香气等，还可以帮助高压工作环境工作人员维持注意力，提高工作效率（I Kei，2015；Sahlim，2014）。

Kaplan et al. pointed out that exposure to natural environments enhances people's cognitive ability to some degree (Kaplan, 1989), and that being in natural environment helps people recover their attention better than in urban environment (Herzog, 1997). Horticultural therapy conducted in natural environment helps the patients concentrate and increase attention (Irvine, 2002; Laumann, 2003). Studies have shown that, compared with short-term HT programs, medium and long-term ones can significantly improve the cognitive functioning of elderly patients with dementia (Sugihra Shiho, 2011). In the case of medium-term, medium-frequency HT programs (e. g. for 4 months, twice a month, 1 hour each time), the participants (patients with Alzheimer's disease) get higher scores in the Mini-mental State Examination (MMSE) after treatment (Jigang Sawa, 2003). After long-term, high-frequency programs (e. g. for 6 months, 3-4 times a month, 30 minutes each time), the MMSE scores of the participants (patients with Alzheimer's) are the same as those at the beginning of the treatment, without further decline, and the scores remain at the same level for more than 3 months after the program ends. Moreover, the memory and memory retrospectivity of patients with severe Alzheimer's show significant improvement after receiving the therapy (Sun Minkai, 2018). More studies have shown that horticultural therapy helps children improve their attention and memory(Taylor, 2001; Kima, 2012; Dai Xing, 2018). In addition, horticultural therapy and plant fragrance can help the staff in high-stress workplaces maintain their attention and improve their efficiency (I Kei, 2015; Sahlim, 2014).

4.3.2　改善生理机能
4.3.2　Improving physiological functioning

　　园艺康健可帮助服务对象锻炼肢体运动能力和心肺、神经功能，从而恢复或提升服务对象的活动能力，进而提高生活质量。诸多研究表明，园艺康健可以增强服务对象的免疫能力（Ulrich，1992；1999）；参加园艺活动可以改善服务对象的心血管功能，身处自然环境中，心跳速率会随之下降（Wichrowski，2005）；园艺康健可以改善中风患者的心血管功能和肢体活动能力（Wichrowski，1998）；养老院开展园艺康健可以帮助老人维持身体活动能力，锻炼已有技能如手部操作的精密活动等，并提高生活质量（Shannon，2010）；参加园艺活动或仅仅身处自然环境之中，就可以缓解慢性疼痛（Szofran，1998；Verra 2012；Diette，2003）；参与园艺康健之后，服务对象的骨骼矿物质含量显著增加（安川绿，2005）；园艺康健花园内散步可以帮助改善服务对象的移动能力（Soderback，2004）；癌症患者通过参加以种植果蔬为主的园艺康健疗程后，其身体活动能力得到了显著提升（Blair，2013）；乳腺癌患者在参加了以森林浴、园艺康健、瑜伽等非药物康健为主的综合治疗之后，其身体活动能力恢复明显（Nakau，2013）。另有研究表明，糖尿病患者在参加园艺康健之后病状得到缓解（Armstrong，2000）。

　　Horticultural therapy can enhance the patients' physical mobility and functioning of heart, lungs and nerves, thus restoring or increasing their physical performance and improving the quality of life. Studies have shown that horticultural therapy enhances the immunity of patients (Ulrich, 1992; 1999). Gardening activities can improve cardiovascular function of the participants and their heart rates slow down in natural environments (Wichrowski, 2005). Horticultural therapy improves cardiovascular function and limb activity of stroke patients (Wichrowski, 1998). Horticultural therapy in nursing homes helps the elderly maintain their physical activity, exercise their remaining skills like sophisticated movement of hands, and thus improves their quality of life (Shannon, 2010). Taking part in gardening activities or simply being in the natural environment can relieve chronic pains (Szofran, 1998; Verra, 2012; Diette, 2003). After the horticultural therapy program, mineral contents increase significantly in the participants' bones (Yaskawa Midori, 2005). A walk in the healing garden intended for horticultural therapy improves mobility of the patient (Soderback, 2004). Physical mobility of cancer patients is significantly enhanced after participating in HT programs featuring fruit and vegetable planting (Blair, 2013). Breast cancer patients show obvious recovery of physical mobility after participating in a combination of non-drug treatments, such as forest baths, horticultural therapy and yoga (Nakau, 2013). Other studies have shown that symptoms of patients with diabetes get relieved after taking part in horticultural therapy (Armstrong, 2000).

4.3.3 促进心理健康
4.3.3 Improving mental health

园艺康健对服务对象的心理健康有显著的正向影响。参加园艺活动可以促进失智老人的感情表达，失智老人对植物的成长往往抱有积极情感并愿意表现出来（Gigliotti，2005）。特别是在看到自己栽种的花卉盛开、食用自己栽种所收获的瓜果蔬菜时，服务对象会自主地表达内心的喜悦心情（田崎史江，2005；黑田利香，2001）。此外，在管理和照料植物的过程中，服务对象也被观察到表现出对收获的预期和向往，增加生存欲（梅田，2001）。参加园艺康健可以降低失智老人的障碍行为发生率，园艺康健栽培管理植物为服务对象提供重塑自信的契机，服务对象的自我尊重得到了显著的提高（Rappe，2007）。参加园艺康健或单纯地置身于自然环境之中，可以缓解心理压力，改善抑郁、焦虑等多种负面情绪（Whitehouse，2001；Elsadek，2019）。甚至仅仅是植物的香味带来的嗅觉刺激，也可以使人减轻压力，进入较为放松的状态（Matsubara，2011）；在充满杉木等木材的香味或其主要挥发性成分的空间内就寝，可以增进睡眠质量（孙旻恺，2019；清水邦义，2018）。

Horticultural therapy has a significant positive effect on mental health of the patients. Engagement in gardening activities can improve emotional expression ability of the elderly with dementia. They tend to have positive feelings about the growth of plants and be willing to show them (Gigliotti, 2005). Especially when seeing the flowers they plant are in full bloom or tasting the fruit and vegetables they grow and harvest (Tasaki Fumie, 2005; Komoda Rika, 2001), they are able to express their inner joys on their own. In addition, while managing and taking care of plants, the patients are also observed to show expectation and yearning for a harvest and an increased desire for survival as well (Umeta, 2001). Horticultural therapy programs can reduce the incidence of impaired behaviors among the old people with dementia. Plant cultivation and management involved in HT programs also create an opportunity for participants to regain confidence, and greatly improve their self-respect (Rappe, 2007). Participating in HT programs or being simply exposed to the natural environment can relieve psychological stress and alleviate depression, anxiety and other negative emotions (Whitehouse, 2001; Elsadek, 2019). Even the olfactory stimulation from the scent of plants alone can relieve stress and help the patient relax (Matsubara, 2011). Sleeping in a space filled with the aroma of wood like *Cunninghamia lanceolata* or its main volatile components can improve sleep quality (Sun Minkai, 2019; Shimizu Kuniyoshi, 2018).

4.3.4 提高社交能力
4.3.4 Improving social skills

园艺康健通常在专业园艺康健师的指导下开展，并且时常多人同时进行，因此可用作社交障碍服务对象的社交能力训练。养老院内失智老人在参加园艺康健后，相互交流增多，变得更愿意参与社交活动，表情也更为丰富(Gigliotti，2005)。拥有园艺或者农业经验的服务对象在园艺康健过程中时常互相交流，并教授其他没有经验的服务对象一些种植、管理技巧，从而促进社交(增谷顺子，2010)。有些园艺康健在实施时特意准备较少的园艺工具或材料，促使参加者之间相互借用，以训练服务对象的社会适应性(铃木みずえ，2004)。在园艺康健过程中，与园艺康健师或其他服务对象的交流，可以帮助稳定服务对象的精神状态，缓和有精神疾病或心理健康问题服务对象的情绪，减少行为障碍的发生频率(长仓寿子，2009)。精神疾病患者在参加园艺康健后，其精神状态改观，并更善于扩展新的人际关系(Barley，2012)。医院等机构的园艺康健花园可以为使用人群提供合适的交流场所，诱发人们开展新的社交活动。

Horticultural therapy is usually carried out under the guidance of professional therapists and participated by a group of people, so it can be used as socializing ability training for patients with communication disorders. After participating in horticultural therapy programs, the mentally retarded elderly in nursing homes interact more with each other, being more willing to participate in social activities and showing more facial expressions (Gigliotti, 2005). During the treatment, participants with gardening or farming experiences tend to communicate frequently with each other and teach other patients planting and management skills, which facilitates social interactions (Ikuga Tomoko, 2010). In some HT programs, fewer gardening tools or materials are provided to prompt communication among participants and improve their social adaptability (Suzuki Mizue, 2004). Communication with the therapist or other participants during the HT treatment process can stabilize the patient's psychological state, ease the mood of those with mental illness or psychological problems, and reduce the incidence of their behavioral disorders (Fudo Hisako, 2009). After receiving horticultural therapy, patients with mental illness show improved psychological conditions and are better at building new relationships (Barley, 2012). HT gardens in hospitals and other institutions can provide users with a suitable place for communication, encouraging them to carry out new social activities.

4.4 展望
4.4 Prospects

园艺康健作为一种非药物干预手段，基于人与自然与生俱来的亲密关系，通过栽种、管理、收获植物或观赏自然景观等活动为不同人群提供放松心情、缓解压力、锻炼社交能力、增强心肺功能、提高免疫力、强健四肢、增强认知能力、减轻抑郁、焦躁等负面情感、缓解疼痛、增加自信心和自我认同感等多方面的功效。同时，在医疗领域，园艺康健不存在副作用和耐药性，无论是作为预防疾病的手段还是治疗疾病的辅助手段，均为极具潜力的替代医疗方案。中国传统文化中的天人合一、中医药文化等都可以用于开发具有我国特色的园艺康健项目。建造实施园艺康健的康复花园(包括园艺康健花园、康健花园、康养花园等)和培育具备专业知识的园艺康健师,则是进一步推广园艺康健的关键。

As a non-drug therapy, horticultural therapy is based on the intimate relationship between man and nature. By growing, managing and harvesting plants or appreciating natural landscapes, it effectively helps different patient populations to relax themselves, relieve stress, develop social skills, enhance cardiopulmonary function and immunity, strengthen limbs, improve cognitive ability, reduce depression, anxiety and other negative emotions, relieve pain, and increase confidence and self-identity. At the same time, in medical field, horticultural therapy produces no side effects and drug resistance, so it is a promising alternative treatment modality, whether as a way of disease prevention or as a supplementary treatment. The concept of Unity of Man and Nature in traditional Chinese culture, together with the traditional Chinese medicine culture, can be employed to develop horticultural therapy with Chinese characteristics. The construction of healing gardens for horticultural therapy (including horticultural therapy gardens, rehabilitative gardens, restorative gardens, etc.) and the training of professional horticultural therapists are the key to the promotion of this treatment modality.

思考题

1. 什么是园艺康健?
2. 简述园艺康健与康复花园设计的关系。
3. 在人体健康照护方法中,园艺康健有哪些优势?

Questions

1. What is horticultural therapy?
2. Briefly describe the correlation between horticultural therapy and healing garden design.
3. What are the advantages of horticultural therapy in health care?

第 **5** 章

康复花园基础设计

Chapter 5　General Design Guidelines for Healing Gardens

健康研究项目和园艺康健项目证实健康与花园之间存在着一定的联系。花园不仅视为物理疗法的源头，而且还是精神疗法的源头。所有人都在谋求健康，并且希望自己拥有一个健康的体魄，因此，健康是人类社会的永恒主题。我们不应该只将康复花园的概念归结为患者的需求，因为康复花园对健康人来说也同样重要(戴维·坎普，2016)。

——摘自海瑟姆·厄尔·巴姆雷博士

"There is a certain connection between health and gardens, which has been confirmed by the Health Research Project and the Horticultural Therapy Project. Gardens are not only regarded as the source of physical therapies, but also the source of psycho ones. People are seeking health and want to be healthy. Therefore, health is the focus of human society all the time. We should not just attribute the concept of gardening therapy to the needs of patients, for it is also important to healthy people (David Kamp, 2016)."

——Excerpt from Dr. Hesham Ei Barmelgy

狭义的康复花园局限于医院、疗养院、精神病院、儿童医院、专科诊所、养老机构等医疗卫生设施的室外庭园景观范围。在这些场所长期或短期居住的人们患有一定疾病或身体较为脆弱，需要经过特殊设计的环境对其进行辅助治疗，以期促进病症康复或达到更好的身心状态。广义的康复花园并非仅仅是指对患者有疗愈功能的花园，也包括所有对人类健康有益的园林绿地。戴维·坎普将康复花园划分为以下三种类型：①自然保护区或者天然水库等天然公园、城市绿化带、绿道、城市空地等绿色基础设施，在这里自然和野生动物能够受到保护；②医院或医疗保健中心内的冥想花园（笔者认为，这一概念可拓展至医疗机构的整个外环境空间，均可纳入康复花园的范畴，而不仅仅指冥想花园）；③私人花园，这一类花园在中国的传统家宅园林里早已有其雏形。

Narrowly, healing gardens are limited to the outdoor landscape in those healthcare facilities such as hospitals, nursing homes and psychiatric hospitals, children's hospitals, specialist clinics, and elderly care institutions. People who have lived in these places for a long or short time must have suffered from certain diseases or are relatively weak, so it is better to treat them in a specifically designed environment in order to enhance in part their recovery or reach a better condition mentally and physically. Broadly, healing gardens not only refer to gardens that can heal patients, but also include all landscapes that are beneficial to human health. David Camp divides healing gardens into three types：①nature parks such as nature reserves or natural reservoirs, and green infrastructures such as urban green belts, green ways and open spaces, where wild animals can be protected；②meditative gardens in hospitals or health care centers (the author believes this concept is too narrow, for all the exterior space of those medical institutions should also be included as a part of healing gardens)；③private gardens, the prototype of them may come from the Chinese traditional gardens built in private houses.

本章将重点介绍目前较为普遍的公共卫生设施康复花园的基础设计指南，希望能够从一般综合性公共卫生设施的室外空间设计入手，扩展到专业性以及针对性较强的、为不同类型群体使用的专类康复花园，如儿童康复花园、阿尔茨海默病疗养花园、老人康复花园等。对于广义范畴的城市绿地内康复景观设计暂且不做讨论。同时将针对康复花园中较为重要的亲生命造园要素——植物与水系进行重点阐述。另外，针对造园工程的概预算也在本章做简要介绍。

This chapter will present a general guideline of how to design healing gardens for those common public health facilities, ranging from the design of outdoor spaces for general public health facilities to those more specialized for different targeted groups, e. g. the healing gardens for children, Alzheimer and old people. The design of healing landscape in urban green spaces in a broader sense is excluded here. At the same time, we will focus on the plants and water system, which are the more important living elements in the healing garden. In addition, the budget for the landscaping project is also introduced in this chapter.

5.1　设计目标
5.1　Goals of design

5.1.1　安全目标
5.1.1　Safety

无论是医疗机构还是普通健康机构的户外公共空间，其设置的目的是让使用者获得健康。无论是针对健康人群还是针对某一方面身心脆弱的人群，设计师在设计花园时首要考虑的就是安全性。在美国，所有的户外空间设计必须遵循健康保险携带责任法案（HIPAA）中要求的保证身心安全的原则，同时，对大部分空间要求符合无障碍——《美国残疾人法案》（*Americans with Disabities Act*）ADA 的通用设计原则。安全目标是康复花园设计最基本的目标之一。

Whether it is an outdoor public space in a medical institution or in an ordinary health institution, its purpose is to make people healthy. Whether it is designed for healthy people or people who are physically sick or mentally vulnerable, safety is the primary factor that should be considered. In the United States, all outdoor space designs must follow the principle of ensuring physical and mental health of people required by Health Insurance Portability and Accountability Act (HIPAA). At the same time, most of the spaces in healing gardens should be accessible under ADA universal principle. Safety is always one of the most basic goals for the design of healing gardens.

5.1.2　情感目标
5.1.2　Emotion

除了安全，花园的所有元素都必须舒适。总体目标是创造一个人们感到被关心的环境。当人们在身体上和情感上都感到舒适的时候，很可能会在花园中停留更长时间，受益也更多。因此，设计要提供安全舒适的地方，以便散步和休憩；同时要为使用者创造选择的机会和对场地的可控性，以便促成人与植物的互动，以及增进社交。这些都会增强身体和情感的舒适度。这样的环境，也会帮助人们从室内陌生、压抑、有威胁的环境中积极逃离，或者暂时忘却自身的疾病和困扰，沉浸在美妙的自然之中。

In addition to safety, all the other elements of healing gardens must make people feel comfortable. The overall goal is to create an environment where people can feel being cared. When people feel comfortable physically and mentally, they are likely to stay in a garden longer and benefit more. Therefore, when designing healing gardens, we would try to provide the safe and comfortable places for people to take a walk and rest, create the feeling that they can choose and control things around them, and provide the opportunities that they can communicate with plants and have more social contact with other people. Such an environment will also help people actively escape from those indoor places, unfamiliar, depressing and threatening, or make them temporarily forget their illness and problems, hence immersing themselves in the beauty of nature.

5.1.3　康复目标
5.1.3　Healing

对于因长期处在压力较大的环境下，导致身体或者心理出现健康问题的人群，以缓解压力、调整情绪、较快地恢复健康为目标，是花园设计的主题。研究表明，与自然接触，尤其是与健康的环境接触，是远离压力环境的最有效方法。与自然接触的程度越深，康复的效果越显著。植物、自然材料、自然界的声音、水景等都是很重要的分散注意力的方法。

For people who have been in stressful environments for a long time and have certain physical or mental health problems, the design of healing gardens is to help them relieve stress, adjust mood and return to a healthy state as soon as possible. Studies have shown that contacting nature, especially with a healthy environment, is the most effective way to stay away from stress. The deeper they contact nature, the more significant the healing effects will be. Plants, natural materials, sound of nature, waterscape, etc. are all important ways of distraction.

5.2　设计原则
5.2　Principles of design

5.2.1　"一体化"原则
5.2.1　Integrative

在做医院及各种健康机构的总体设计时，应采取建筑和户外空间"一体化"的规划设计思路，将其作为一个利于健康的整体环境进行考虑。即户外空间并非建筑无法使用残留的灰空间，而是要求总体规划师能够在做建筑规划时，充分考虑户外空间的位置、大小、形态，以及与建筑的连接关系等方面，使户外空间的使用能够融入机构的日常生活。

It is recommended that hospitals and various health institutions adopt an "integrated" planning and design approach for buildings and outdoor spaces in the overall design of a healing garden. In other words, planners should fully consider the location, the size, the shape of outdoor spaces as well as the connection with buildings, so as to integrate the space utilities into the institution routines. It means that the outdoor spaces would not be seen as the ones of little value that have to be kept with the main buildings.

5.2.2　可达性原则
5.2.2　Accessible

可达性是花园设计的基本原则。可达性包含两个方面：一是视觉上的可见，二是身体上的可达。医院或者健康机构内设置的花园首先要"可见"，如果过于隐蔽，且缺乏导视系统，那么使用者就很难发现并使用花园。因此，要将花园布置在人流相对集中、且潜在使用者可以目及之处；如果无法做到"可见"，那么就需要规划合理的路径、设置清晰的导视系统，让使用者知道花园的位置、大小、抵达路径、何人可以使用花园、花园内可以开展什么活动等内容。身体的可达，首先需要保证从建筑进入花园的门方便开启；其次考虑在花园入口处提供一个平缓的无障碍区；最后要尽可能保证花园的门随时开放，如果需要在特定时间关闭，则需要在花园入口处做好说明等。

Accessibility is a basic principle of designing healing gardens, which includes the following two aspects: one is visual visibility, the other is physical accessibility. The healing garden in a hospital or a health institution must be designed to be "visible". If it is too concealed and lacks a guiding system, the garden will be difficult to be found. Therefore, it is best to set up the garden in a place where the flow of people is dense so that it can be easily seen by the potential clients; if it cannot be "visible", a clear guidance system is necessary. By the guidance, people can know the location and the size of the garden, the ways to get there, who can enjoy the garden, and what activities would be carried out in the garden, etc. The healing garden must be physically accessible. It means that first of all the entrance door from the main building to the garden should be easily opened. Besides, a flat barrier-free area should be provided at the entrance of the garden. And then, ensure that the door to the garden is open all the time. If the door needs to be closed at a specific time, a notice should be given in time at the entrance.

5.2.3　操控性原则
5.2.3　Controllable

研究表明，康复花园操控感的缺失，可能导致或者加重花园使用者的沮丧和消极情绪，或进而导致血压升高、免疫力下降等不良后果。让使用者参与花园的设计中，可以提升使用者对花园的操控感，使用者可以对不同空间、布局，甚至休息设施进行选择，从而实现对所处空间的掌控，增强自信心。

Studies have shown that sense of being uncontrollable may cause people feel depressive and even make it worse, and increase their blood pressure and decrease their immunity. Allowing the clients to participate in the design of a healing garden can enhance their sense of control over it. They can make a choice of different spaces and layouts or rest facilities by themselves, so as to feel that they can control the environment and enhance self-confidence.

5.2.4　安全性原则
5.2.4　Secure

患者的身体和心理都十分脆弱，因此，应该给予患者更多的安全感。为了达到这一目的，需要从花园规划设计的多个方面加以考虑。例如，保证花园内有充足的照明设置，并在偏僻的区域设置公用电话以方便患者求救；道路铺装设计尽量选择防滑、安全的材料，适合不同类患者的使用；植物材料的选择不仅要考虑美观，而且要充分考虑花粉等刺激物对患者的影响等。

Clients and patients should be given more sense of security for they are both physically and mentally weak. In order to achieve this goal, many aspects need to be considered in planning and designing, e.g. adequate lighting, public phones set up in remote districts to facilitate patients to call for help, non-slip and safe materials suitable for different types of patients for road paving. Of course, the choice of plants is not only for attraction, but also for fully considering the impact of pollen and other stimuli on the patients.

5.2.5　舒适性原则
5.2.5　**Comfortable**

花园设计要充分考虑患者对温度和阳光的敏感性。例如，烧伤患者需要避免阳光直射，因此，需要在花园内设置光照区和庇荫区，以及遮阴座椅。多种药物疗法均需要患者避免阳光直射；座椅的设置最好带有扶手和靠背，提供可供平躺的躺椅，为站立困难的患者提供方便；坐凳面选择木质材料，满足北方地区不同季节使用座椅的舒适性。同时，对不同类型的使用群体，分别考虑其对私密性的要求，设置一些满足个人静思或者两三人聊天、交流的区域，也是非常必要的。

Also, the design of a healing garden should take into consideration patients' sensitivity to temperature and sunlight. For example, the burn patients need to avoid direct sunlight, so it is necessary to set up both sunlight and shade areas in the garden, as well as the seats that are shaded by plants and building structures. In fact, multi-drug therapies require the patients to avoid direct sunlight. Moreover, it is better to set up a seat with armrests and backrests, provide a recliner that can lie flat for those with difficulty in standing, and choose wooden materials in order for the seat's comfort in different seasons in northern areas. At the same time, it is necessary to set up relatively private areas for the patients to meditate alone or chat by two or three.

5.2.6　可持续性原则
5.2.6　**Sustainable**

花园各种设施系统和景观系统要充分考虑耐久性和可持续性，以便长时间内保持良好的景观效果和使用。例如，雨水系统一方面可以采用透水的铺装材料，另一方面可以结合景观设置一定数量的雨水收集绿地，进行生态雨水的收集和回用；植物配置设计充分考虑植物生长变化的过程，以及不同类型的植物对雨水光照等方面的要求，尽可能选择多年生植物代替观赏草坪等，若需要设计草坪也应选择耐践踏草种，设计为游憩型草坪，总之是要减少绿篱、草坪等的维护成本，同时，营造丰富的季相变化。

Durability and sustainability of various facility systems and landscape systems in a healing garden should also be considered in order to keep their service longer and maintain better effect. For example, on the one hand permeable paving materials should be applied in rainwater system; on the other hand, a number of rainwater collection green spaces can be set up to collect and reuse ecological rainwater. As to plant allocation, the process of plant growth and changes as well as the requirements of sunlight and rainwater for different plant types should be comprehensively considered. Choose perennial plants instead of ornamental lawns, for instance. The lawns, if necessary, should be designed as recreational by using tramp-resistant grasses. In short, it is necessary to reduce maintenance costs of hedgerows and lawns and present rich seasonal changes of them at the same time.

5.2.7 参与式设计原则
5.2.7 **Participatory**

与普通花园设计相比，康复花园需要从顶层设计开始就建立参与式的设计流程（PDP）。通常情况下，设计项目分为"甲方"和"乙方"，整个设计流程是一个双边关系；而参与式设计流程需要由一个交叉学科组成的团队来执行，比如拥有"投资方""研究方""设计方""使用方"，这是一个多边关系的团队，将康复训练专家或研究者与设计师和建筑师整合到一起，患者、家庭成员和志愿者也作为团队的成员，大家共同围绕一个既定的康复花园设计目标进行方案的讨论与定夺，包括后期的施工建造过程，都将由该团队全程引领。参与式设计流程的关键，就是将了解患者临床需求的医疗团队成员聚集在一起，并将他们与员工、志愿者和家庭的需求结合起来。这种花园设计方法不仅提供了一个美丽的环境，而且具有满足不同群体需求的功能，最重要的是使后期花园的使用和维护获得可持续的保障。

Compared with an ordinary garden, a healing one needs to establish a participatory design process（PDP）at the beginning. Commonly, the parties in an ordinary design are divided into "Party A" and "Party B". The entire design process is just bilateral. While PDP needs an interdisciplinary team which may include the parties of "investor" "researcher" "designer" and "user". This team is a multilateral one that makes healing training experts, researchers, designers and architects get together. More importantly, patients as well as their family members and volunteers have become a part of the team. Everyone can discuss and decide the plan around a specific goal of designing a healing garden, including the later construction process. The key to PDP is to bring together medical team members who understand the clinical needs of patients and integrate them with employees, volunteers and families. This method not only provides a nice environment, but also plays the role of satisfying the needs of these different groups. The most important, for doing so, is that the garden's usage and maintenance turn to be sustainable.

5.3 亲生命的设计要素
5.3 **Living elements of design**

在康复花园的造园要素中，植物和水可谓亲生命的设计要素，本节重点介绍集保健与观赏功能于一身的药用植物资源和水系组织。

Among the elements of healing garden design, plants and water are living elements. The medicinal flower resources and water system organization integrating health care and ornamental functions will be introduced in this section.

5.3.1　康复花园药用花卉资源
5.3.1　Medicinal flower resources of healing gardens

"原本山川，极命草木"，草木筑品，为君康健。同是大自然生灵的植物与人在漫长的历史进化中，形成了极为密切的关系。植物的花朵是神秘的、美丽的、生机勃勃的，可入诗、可入画、可入药，甚至可以与人们的心灵相印。直接用花来医病，已有相当长的历史。我国的处方药里，有 1/4 含有开花植物的某部位或其合成物，目前开发出药用疗效的开花植物仅约占全球植物的 1%，所以这一数据仍然有继续增长的潜力。花卉与人类健康的关系，除了入药直接治疗人体疾病以外，更能从思想和精神的层面让人感到宁静和谐。

An ancient Chinese poem called for the scholars "to explore the geographical conditions of mountains and rivers, and to strive for exhaustive researches of plants". Planted-based creations can contribute to human well-being. Plants and humans have formed a close relationship during the long historical evolution. The flowers of plants are mysterious and magnificent. They are described in poems, drawn in paintings, used as medicines, and more importantly, implanted in our souls. It is a long history of directly using flowers to cure diseases. There are a quarter of prescription drugs which contain a certain part of a flowering plant or its compositions. However, the flowering plants that have proved to be medicinally effective currently account for only about one percent of the global plants. So these data will surely be growing in future. Flowers can not only be used as medicine to directly cure diseases, but also make people feel peaceful and become harmonious physically and mentally.

5.3.1.1　药用花卉概述
5.3.1.1　Overview of medicinal flowers

药用花卉是指在人们的应用中，以其药用价值为主，观赏价值为辅的花卉。人们熟悉的有防病治病、保健强身、延年益寿作用的许多中草药，其原植物也是花姿秀丽、枝叶美观、果形悦目、气味芳香的花卉。所以，它们在作为药材栽培的同时，也常用于装饰环境和供人观赏。尤其是我国人民千年来在药用花卉实践中积累的经验与智慧，为当今康复花园的植物选择提供了丰富又宝贵的资源。

Medicinal flowers refer to those that are used by humans mainly for medicinal purposes with ornamental value as a supplement. People are familiar with many Chinese herbal medicines that have the effects of preventing and curing diseases, strengthening body and prolonging life. These herbs are originally flower plants with lush branches and leaves, pleasing fruits and fragrant smell. Therefore, while they are cultivated as medicinal plants they are also used to decorate environment and for people to view and enjoy. For thousands of years Chinese people have accumulated experience and wisdom in the practice of adopting medicinal flowers, and hence provided rich resources for the selection of plants in present healing gardens.

据《神农本草经》记载：菊、百合、芍药、牡丹、鸢尾、射干、瞿麦、连翘和辛夷等，均是著名的药用观赏花木；《本草纲目》也记述了近千种花卉的性状、功能和主治疾病，如茉莉、迎春、蜡梅等；《中华人民共和国药典（2010 版）》收录了鸡冠花、金银花、月季、凌霄等花卉。而民间惯用的药用花卉，如仙人掌、玉簪、君子兰等，在防病治病中的应用就更为广泛（周武忠，1999）。近百年来，对花卉的药用研究发展迅速，已发现了很多花卉有着极其显著和广泛的药理作用。随着药理作用和化学成分研究的深入，对不少种类花卉已分析出了其有效成分和有效部位，临床应用范围逐渐扩大，并取得了新的应用成果。如长春花有治疗糖尿病的功效；该植物的萃取物还可以降低白细胞数，抑制骨髓的活动，通过试验发现，两种源自长春花的化学物质在治疗儿童白血病时，病童的存活率由 10% 增至 95%（萝赛，2004）。人们周围的不少花卉有治疗效果：如丝兰是类固醇；毛蕊花是温和的镇静剂，其根可增加膀胱的张力，避免尿失禁；杜松可以治疗膀胱炎；当归、矢车菊、月见草、甘草、益母草、欧薄荷、牡丹、覆盆子等可以治疗月经阵痛；锦葵、委陵菜、鼠尾草等对扁桃体炎有疗效；吊钟柳和蓟罂粟抗晒伤（张春晓，1996）。

It is recorded in *Divine Farmer's Classic of Materia Medica*: Chrysanthemum, Lily, Peony, Tree Peony, Iris, Belamcanda, Dianthus, Forsythia, and Magnolia are all well-known medicinal and ornamental plants. *Compendium of Materia Medica* also describes the appearance, the nature and the function of nearly a thousand flowers (e. g. Jasmine, Winter Jasmine, Wintersweet, etc.) as well as the diseases that they can help to cure; *Chinese Pharmacopoeia* published in 2010 has included Cockscomb, Honeysuckle, Chinese Rose, Campsis Grandiflora and other flowers. The folk medicinal flowers, such as Cactus, Plantain Lily, and Clivia, are more widely used in disease prevention and treatment (Zhou, 1999). In the past hundreds of years, medicinal research of flowers has developed so rapidly that many flowers have been found to have significantly extensive pharmacological effects. With the in-depth clinical study of pharmacological effects and chemical components, many kinds of flowers have been experimented as to their ingredients and parts that may be effective in curing diseases. For example, Periwinkle is proved to be a good remedy for diabetes, for its extracts can reduce the count of white blood cells and inhibit the activity of bone marrow. Experiments showed that two chemicals derived from the Periwinkle, when used in treating childhood leukemia, can increase the survival rate of the sick children from 10% to 95%(Luo, 2004). Almost all flowers around us have therapy functions: Yucca is a steroid; Mullein is a mild tranquilizer, and its roots can increase the tension of bladder and prevent urinary incontinence; Juniper can treat cystitis; Angelica, Cornflower and Scabish, Liquorice, Leonurus and Peppermint, Tree Peony, Raspberry, etc. can treat menstrual pain; Mallow, Potentilla, Sage, etc. are effective for amygdalitis; Penstemon and Thorn-poppy can be used against sunburn (Zhang, 1996).

5.3.1.2　中国传统十大名花药用价值

5.3.1.2　Medicinal value of Chinese Top Ten Traditional Famous Flowers

中国传统十大名花不仅从色香姿韵给人以精神上的美感，同时也能从衣食住行方面直接为人类提供物质资源。就药用价值而言，中国传统十大名花针对不同疾患都有其各自的疗效，表 5-1 按花名拼音排序列出其药用功能（郭文场，1989）。更多详细内容见附录 2。

Chinese Top Ten Traditional Famous Flowers not only give people a spiritual enjoyment by their colors, fragrances and shapes, but also directly provide them with daily life resources. In terms of medicinal value, Chinese Top Ten Traditional Famous Flowers respectively have their curative effects for different diseases. Tab. 5-1 lists the medicinal functions of Chinese Top Ten Traditional Famous Flowers in alphabetical order by their Pinyin names (Guo, 1989). For more detailed information, please refer to Appendix 2.

表 5-1　中国传统十大名花药用功能一览表（按中文首字母拼音排序）

植物名称	拉丁学名	作用及功效
杜鹃花	*Rhododendron simsii*	清热解毒，化痰止咳，止痒，祛风湿，活血化瘀
桂　花	*Osmanthus fragrans*	散寒破结，化痰止咳，暖胃平肝，祛风湿，散寒
荷　花	*Nelumbo nucifera*	凉血散瘀，补心益胃，益肾固精，清热利湿，解暑除烦
菊　花	*Chrysanthemum morifolium*	疏风散热，养肝明目，解疔疮毒
兰　花	*Cymbidium*	滋阴清肺，化痰止咳
梅　花	*Prunus mume*	敛肺涩肠，除烦，生津止渴，杀蛔，止血，活血解毒，平肝胃
牡　丹	*Paeonia suffruticosa*	清热凉血，活血化瘀（制炭用于止血）
山　茶	*Camellia*	收敛凉血，止血，调胃，理气
水　仙	*Narcissus tazetta* subsp. *chinensis*	清热解毒，散结消肿，活血通经
月　季	*Rosa chinensis*	活血调经，消肿解毒

Tab. 5-1　A list of Medicinal Values of the Chinese Top Ten Traditional Famous Flowers
（According to Their Initials of Chinese Names by Pinyin Alphabetic Order）

Plant name	Latin name	Function and efficacy
Azalea	*Rhododendron simsii*	Clear internal heat and remove toxicity, eliminate phlegm and stop cough, and relieve itch, expel internal dampness, stanch bleeding
Sweet Osmanthus	*Osmanthus fragrans*	Dispel internal cold and eliminate coagulation, eliminate phlegm and stop cough, warm the stomach, calm the liver and dispel internal cold

（Continued）

Plant name	Latin name	Function and efficacy
Lotus	*Nelumbo nucifera*	Cool the blood and dissipate stasis, nourish the heart and benefit the stomach, tonify the kidney and reinforce essence, clear heat and eliminate dampness, relieve summer heat and dispel annoyance
Chrysanthemum	*Chrysanthemum morifolium*	Expel internal dampness or heat, nourish the liver and improve Eyesight, relieve poison of skin ulcer
Orchid	*Cymbidium*	Enrich Yin* system and clear away lung-heat, eliminate phlegm and stop cough
Plum	*Prunus mume*	Astringe the lungs and intestines, relieve restlessness, help produce saliva and quench thirst, kill roundworms and stanch bleeding. promote blood circulation and remove toxicity, calm the liver and the stomach
Tree Peony	*Paeonia suffruticosa*	Remove pathogenic heat from blood, promote blood circulation to remove stasis (it can be used to make carbon to stanch bleeding)
Camellia	*Camellia*	Astringe and remove pathogenic heat from blood, stanch bleeding, nourish the stomach and regulate the flow of qi**
Narcissus	*Narcissus tazetta* subsp. *chinensis*	Clear internal heat and remove toxicity, eliminate coagulation and disperse a swelling, promote blood circulation and make menstrual flow smoothly
Chinese Rose	*Rosa chinensis*	Promote blood circulation for regulate menstruation, disperse a swelling and remove toxicity

5.3.1.3　常用花草茶植物药用功能简介

5.3.1.3　Brief introduction of medicinal values of commonly used herb tea plants

花草茶是将植物的花叶或果实，经过一定的晾晒或炮制以后制成的茶饮。花草茶的制作简单有趣，常用于康健园艺活动中打开味觉、嗅觉、触觉通道的实践和体验。花草茶在养护脏器、解毒排瘀、降压降脂等方面具有功效。表5-2列出了几种常用的花草茶植物及功能。

表5-2　常用花草茶植物一览表

植物名称	拉丁学名	作用及功效
马鞭草	*Verbena officinalis*	全草药用，活血散瘀，解毒，利水消肿
迷迭香	*Rosmarinus officinalis*	消除胃气胀，增强记忆力，提神醒脑，减轻头痛症状，调理油腻不洁的肌肤，促进血液循环，改善脱发
柠檬草	*Cymbopogon citratus*	利尿，防止贫血及滋润皮肤，健脾健胃

---

* Yin and Yang, opposite to each other, are two important concepts in the science of traditional Chinese medicine. Yin mainly refers to the fluids in the body, including blood, saliva and tears, endocrine, oil ester secretion, etc. ; while Yang refers to the mechanisms of body. The core of traditional Chinese medicine is to try to keep balance between Yin and Yang.

** It refers to the vital energy of body.

（续）

植物名称	拉丁学名	作用及功效
洋甘菊	*Matricaria chamomilla*	治疗失眠，降低血压，增强活力、提神，增强记忆力、降低胆固醇、纾解眼睛疲劳
玫瑰茄（洛神花）	*Hibiscus sabdariffa*	调节和平衡血脂，增进钙质吸收，促进儿童发育，促进消化等
甘 草	*Glycyrrhiza uralensis*	清热解毒，祛痰止咳、温腹
枸 杞	*Lycium chinense*	补肝益肾，调节机体免疫功能，有效抑制肿瘤生长和细胞突变，延缓衰老，调节血脂和血糖，促进造血
藿 香	*Agastache rugosa*	祛暑解表，化湿和胃

Herb tea is made from the floral leaves or fruits of plants after drying or processing. The production of herb tea is simple and interesting, and it is often used in the activities of healing gardens to help clients stimulate their senses of taste, smell and touch. It is generally believed that herb tea has the effects of protecting organs, detoxifying and removing blood stasis, lowering blood pressure and lipids, etc. Tab. 5-2 lists several commonly used herb tea plants as well as their efficacy.

Tab. 5-2　A List of Commonly Used Herb Tea Plants

Plant Name	Latin Name	Function and Efficacy
Vervain	*Verbena officinalis*	This plant can be wholly used as medicine. It can promote blood circulation to remove stasis, detoxify, help urinate and disperse a swelling
Rosemary	*Rosmarinus officinalis*	This plant can release stomach bloating, enhance memory and make refreshing, relieve headache, condition oily skin, promote blood circulation and improve hair loss
Lemongrass	*Cymbopogon citratus*	This plant can nourish the spleen and the stomach, help urinate, prevent anemia and moisturize skin
Chamomile	*Matricaria chamomilla*	This plant can treat insomnia, lower blood pressure, enhance vitality, make refreshing, enhance memory, lower cholesterol and relieve eye fatigue
Roselle	*Hibiscus sabdariffa*	This plant can regulate and balance blood lipids, increase calcium absorption, promote children's growth and promote digestion, etc
Licorice	*Glycyrrhiza uralensis*	This plant can clear internal heat and remove toxicity, eliminate phlegm and stop cough, and warm the stomach
Wolfberry	*Lycium chinense*	This plant can invigorate the liver and the kidney, regulate immune function, inhibit tumor growth and cell mutation, delay aging, regulate blood lipids and blood glucose, and promote hematopoiesis
Wrinkled Giant Hyssop	*Agastache rugosa*	This plant can dispel the heat of summer to relieve superficial syndrome, dissipate hygrosis to warm the stomach

5.3.1.4　常用神经系统病症用花

5.3.1.4　Flowers commonly used for symptoms of nervous system

　　康复花园对于缓解神经系统常见病症诸如神经衰弱、健忘多梦、偏头痛、平日多忧愁、忧郁烦闷、心神不安、心悸失眠等多有帮助，表5-3列出了我国医学中对这些病症常用的花疗用花，以供康复花园的植物种植设计参考选用。部分花卉功效详见附录3。

表5-3　常用神经系统病症用花

神经系统病症	花疗用花
神经衰弱	百合、含羞草、梅花、茑萝、兰花
健忘多梦	合欢花
偏头痛	梅花、向日葵
平日忧愁	萱草
忧郁、烦闷	金橘
心神不安	合欢花、景天
心悸、烦躁	景天
失眠	含羞草、荷花、荷包牡丹、景天、萱草

Healing gardens are helpful for alleviating the symptoms of common nervous system such as neurasthenia, forgetfulness and dreaminess, daily annoyance and worry, migraine, moping and depression, restlessness, palpitation and insomnia, etc. Tab. 5-3 below lists the flowers commonly used in Chinese medicine for these diseases. This can be a reference for the design of plants growth in healing gardens. The efficacy of some flowers can be found in Appendix 3.

Tab. 5-3　Flowers Commonly Used for Symptoms of Nervous System

Nervous System Symptom	Flowers Used in Therapy
Neurasthenia	Lily, Mimosa, Plum, Cypress Vine, Orchid
Forgetfulness and Dreaminess	Silk Tree
Migraine	Plum, Sunflower
Daily Annoyance and Worry	Daylily
Moping and Depression	Kumquat
Restlessness	Silk Tree, Sedum
Palpitation and Irritability	Sedum
Insomnia	Mimosa, Lotus, Sealflower, Sedum, Daylily

　　除了以上入药花卉以外，中国丰富的花卉资源中还有大量可供药用的花卉（刘小勤，2005）。张春晓（1996）在《花卉栽培与药用》一书中收录了121种常见的药用花卉及其疗效，并对应各系统常见332例病症列出了可治疗该疾病的各种花卉。因篇幅所

限，本节按照药效功能及主治范围，将其最为常用的药用花卉归纳如下，以供参考。

In addition to the medicinal flowers mentioned above, there are a large number of others that can be used as medicine in overall China's flower resources（Liu，2005）. Zhang Chunxiao（1996）included 121 common medicinal flowers and their effects in *Flower Cultivation as well as Medicinal Use* and listed various flowers that could treat 332 common cases for different human systems. For the sake of limited space, this section summarizes part of the most commonly used medicinal flowers in accordance with their medicinal functions and scope of curing.

①清热解毒类　鸡冠花、金莲花、紫薇花、木棉花、木芙蓉、木槿花、茉莉花、栀子花、金银花、凌霄花、美人蕉、百合花、萱草、玉簪花、仙人掌等。

①Clearing internal heat and detoxifying　Cockscomb, Nasturtium, Crape Myrtle, Cotton Tree Flower, Hibiscus Mutabilis, Cottonrose Hibiscus, Jasmine, Gardenia, Honeysuckle, Campsis Grandiflora, Canna, Lily, Daylily, Plantain Lily, Cactus, etc.

②舒经活血类　八仙花、紫茉莉、叶子花、秋海棠、垂丝海棠、紫荆花、迎春花、金盏菊、文竹、一叶兰、珠兰、紫色鸭跖草、景天等。

②Making menstrual flow smoothly and promoting blood circulation　Hydrangea, *Mirabilis jalapa*, Bougainvillea, Begonia, *Malus halliana*, Bauhinia, Winter Jasmine, Calendula, Asparagus Fern, *Aspidistra elatior*, Chloranthus, Purple Dayflower, Sedum, etc.

③止咳化痰类　白兰花、蜡梅花、桂花、夹竹桃、昙花、石楠花、桔梗、桃花、柚子、向日葵、虞美人、苏铁、芦荟、万年青等。

③Eliminating phlegm and stopping cough　Gardenia, Wintersweet Flower, Osmanthus, Oleander, Epiphyllum, Moor Besom, *Platycodon grandiflorum*, Peach Blossom, Grapefruit, Sunflower, Corn Poppy, Cycad, Aloe, *Rohdea japonica*, etc.

④止疼镇痛类　玉兰花、芍药花、凤仙花、瑞香、女贞、枸杞、黄杨、棕榈、凤仙、鸢尾、天门冬等。

④Relieving and easing pains　Magnolia Flower, Peony Flower, Garden Balsam, Winter Daphne, *Fructus ligustri*, Wolfberry, Boxwood, Palm, Impatiens, Iris, Radix Asparagi, etc.

5.3.2　康复花园水系组织
5.3.2　Healing gardens' water system

5.3.2.1　水景与健康
5.3.2.1　Waterscape and health

水是生命之源，人类择水而居，可见对于水的依赖。水在日常生活中，除了满足人们的生存需求以外，作为花园景观中不可或缺的美，被人们赋予人文色彩和精神寄托。正如唐纳德·牛顿·韦伯所说"有水的地方就有花园"。花园中的水系，具有丰富园景构成、美化空间结构、软化景观机理等视觉作用。此外，苏黎世大学的研究表明："水流的声音有助于预防及减缓压力的堆积。"同时，水也能够释放有利于身体健康的空气负离子，使人头脑清新、呼吸舒畅。因此，水既可以创造生命，维护生命，有益于健康，又可以装点环境，给人们以美的享受。

Water is the source of life. Human beings choose to live near water, which shows their dependence on water. Water in daily life, in addition to satisfying people's survival, is endowed with the color of humanism to become their spiritual dependence. It is an indispensable beauty in a garden scene. As Donald Newton Weber said, "Where there is water, there is a garden." The water system in the garden can visually enrich the landscape composition, beautify the spatial structure and soften the mechanism of the landscape. In addition, a study by the University of Zurich showed that the sound of water flow helps prevent and relieve stress. At the same time, water can also release air anionic substances that are beneficial to health, which can refresh people's minds and make them breathe comfortably. Therefore, water can not only create and maintain life, beneficial to health, but also decorate environment and give people better enjoyment.

5.3.2.2　康复花园的水景
5.3.2.2　Waterscape in a healing garden

康复花园以人工和自然的介入为手段，让人有安全感，减少压力，增强舒适和活力。花园内应该有一定程度/范围的绿色植物、花、水等元素，它们能为大多数使用者提供治疗和助益。排除特殊不适宜人群外，作为以康复为目的的水景，其特点及形式与园林设计中的一般水体及周边组织的要求基本相同。此外，水体及其质量，决定着人们的生存质量，同时在城市软实力构建中，也是一项重要指标。可见，人与水的关系相辅相成密不可分。

A healing garden is a place where people can feel safe, less pressed, more comfortable and energetic through both artificial and natural interventions. There should be a certain amount of green plants, flowers, waterscapes, etc. in the garden, which can help treat and benefit most clients. As waterscapes for rehabilitation, their characteristics and forms are basically the same as those of the water bodies as well as their surrounding structures required in a general garden design. In addition, water body system as well as its quality, as an important indicator of urban soft power, determines the quality of people's lives. It can be seen that the relationship between man and water is complementary and inseparable.

(1)水景的特质
(1) Characteristics of waterscape

我国著名的水景不胜枚举，有水流潺湲曲折多变的南京瞻园，有明媚如镜的杭州鉴止水园，有喷吐作声的济南趵突泉，还有水声叮咚的无锡八音涧等。水体本身和其边界组织是构成美的基础，通过水的形状、流水方式、声音以及镜像等来表现水景的特征，让人产生美好的联想，这也便是设计康复花园水景的目标。

There are countless famous waterscapes in China, including Nanjing Prospect Park characterized by turbulent waters, Hangzhou Jianzhishui Park in which water is as bright as a mirror, Baotu Spring in Jinan characterized by spurting water, and Bayin Gully in Wuxi with particular knocking sound of water, etc. The water body itself and its boundary surroundings are the foundation of waterscape. The characteristics of waterscape are reflected by the shape, the sound, the flowing way and the mirror image of water, which can give rise to good imagination of people. This is also the goal of designing waterscapes in a healing garden.

①形 水作为中国古典园林的四要素之一，具有可变性和自由性。边界的围合可以形成各种形状的空间对比和交融。园林中水多用拟态的手法，对自然界的江、河、湖、海进行艺术加工后以溪、涧、潭、瀑、池的方式表现。水面聚在一起时给人开阔明朗的印象，而分出的支流小溪则显得蜿蜒曲折，给人以幽静的印象。水面空间的开合变化、水岸的虚实曲直处理，营造了变化万千的水形和水态。

①Formation   As one of the five elements of Chinese classical gardens, water is variable and free. The enclosure of boundaries can create various spatial contrasts and fusions in shape. In garden design, water is often represented through mimetic techniques. The natural lakes, rivers and seas are often artistically imitated in the form of streams, gullies, pools, waterfalls and ponds in waterscapes. When all the water surfaces gather together, the whole waterscapes look wide and bright, while the winding streams that branch out make it appear to be peacefully quiet. The opening and closing of the water surface have created a multitude of diverse shapes and states.

②动与静 园林里水有动态和静态之分，即动水和静水。动水主要指的是园林中的溪流、喷泉、瀑布、跌水、涌泉等，来体现园林水景的动态美。在园林中也有以跌落与喷涌等形式，通过"喷、涌、注、流、滴"等状态塑造出生动的园林环境。静水主要是湖、池、塘、潭等，通过如镜的水面，映出周围的湖光山色。

②Dynamic state and static state   There are dynamic and static water in a garden. Dynamic water mainly refers to streams, fountains, waterfalls, cascades, springs, etc. all of which reflect the dynamic beauty of the garden waterscape. Besides, different water forms such as "spraying, gushing, pouring, flowing, dripping" can shape a lively garden environment. Static water mainly refers to lakes and ponds, which reflect the landscapes all around by their calm mirror-like water surfaces.

③声 水流尽管有"声"，却能让人融入自然、带来宁静平和的感觉。利用水体营造声景，凭借地形的高低及山石等材质，设计引导出水流的声音。从跌水及瀑布的落水声响中表现出或急或缓、或清脆或暗哑等变化。花园是个整体，就像水在环境中并不是单独存在一样，有水就有山、有植物、有人等。因此声音的设计也要考虑到周围声环境，一般除了水流以外，还可以利用风、雨等外界的声音。

③Sound   Although it has "sound", water can make people integrate into nature and feel tranquil and peaceful. Water body would create various sounds relying on different heights of terrains and different properties of hills or stones, etc. For instance, the sound of cascades and waterfalls may alternate rapidly or slowly, crisply or dully. The garden is a whole because water has integrated into the environment. Where there is water, there are mountains, plants, people, etc. Therefore, the design of water sound should also consider the surroundings. In addition to water flow, other external sounds such as wind and rain should also be included.

④镜　利用水的映射效果，把四周景色融入水中，岸边的山体、桥石、建筑等在水中形成倒影，使园景变得宽广而深远。在面积有限的空间，为了弥补场地带来的视觉阻塞及限制，可用较大的水面来扩展视域感。北宋刘攽《雨后池上》所作"一雨池塘水面平，淡磨明镜照檐楹。东风忽起垂杨舞，更作荷心万点声"，以静显动，又以动衬静，动静结合，有镜的映射、有对声的描述，充分表现出水景的特征。使人们看到了自然界自身律动的美，感到亲切，又正是这种亲切感，实现了物境和心境的交融与和谐。

④Mirror effect　Mountains, bridge stones, buildings, etc. on the shore of water can be reflected due to water's mirror effect, which may make views of landscapes deeper and broader. Actually, in order to compensate for visual restriction brought by the site within a limited space, a larger water surface is a good choice to expand the sense of sight. Liu Ban in the Northern Song Dynasty ever wrote in his poem *A Pond after Rain*: "After rain the pond was full of water. It looks so calm like a lightly polished mirror to reflect houses on the shore. Suddenly blows an east wind, gracefully dance the swaying willows, even on the middle of lotus leaves drip the dews from willows." This poem presents a typical water scenery in which the scenes are alternately static and dynamic, with a description of the sound and the reflection of water. It makes people see the beauty of nature and feel intimate with it. It is this intimacy that realizes a harmony between physical form and mental state.

（2）水体在康复花园中的功能与作用

（2）Functions and roles of waterscape

对康复花园而言，水景的设置是非常必要的，它丰富了空间层次，也促使周边的生态系统完整与生动，实现物我同一、相与共生的目的。此项可归纳为以下几点：

For a healing garden, the setting of waterscape is very necessary. It can enrich spatial hierarchy and also make surrounding ecosystem complete and vivid, hence reaching an effect of integrating humans into nature. This can be summarized as follows:

①空间的连续、分隔和拓展　在康复花园内导入水体，可以做到景观相连，互为作用，成为统一整体。水景的引入，能够起到分隔空间的作用，使观者沿着水边绕行，提升审美情趣；水面投入的倒影，还可丰富视觉享受，增加了园景的层次感。

①Continuity, separation and expansion of space　The introduction of water bodies into a healing garden can connect different landscapes and make them interact with each other to become unified. The introduction of waterscape can play a role in separating space, allowing tourists to walk along the river or the lake and enhancing their aesthetic tastes; the reflection of water surface can also enrich tourists' visual enjoyment and increase the dimensional effect of landscapes.

中国传统园林中追求以小见大效应，以有限的空间，追求无限的意境，来拓展空间。

In traditional Chinese gardens, we try to pursue an infinite artistic scene with limited spaces, hence the expansion of the spaces.

②生态系统的保全　哪里有水，哪里就有生命。水体内及周边为生物提供生存和栖息的空间，滋养了水岸的水生动物与植物。所以构建水景观，对于康复花园生物的多样组合及景观体验将起到积极的审美作用。此外，康复花园中，水体在微气候调节中也发挥着重要作用。

②Preservation of ecosystem　Where there is water, there is life. Water body and its surrounding areas provide habitats for organisms, nourish aquatic animals and plants living alongside the water. Therefore, the water landscape plays an aesthetic role for visitors in enjoying the diverse organisms and landscapes in a healing garden. In addition, the water body also plays an important role in regulating microclimates.

③浇灌与防火防灾　康复花园中的水体构建，还有其他作用，如浇灌及消防等，这样不仅可以减少管理成本，也可以有效防止意外发生。

③Irrigation and fire prevention　The construction of water body in a healing garden has other roles, e. g. irrigating and fire-fighting, etc. which can not only reduce management costs, but also effectively prevent accidents.

（3）水系的场地规划

（3）Site planning of water system

水景的形成和场地的自然风貌、地域文化、规划用途及使用对象有关。合理规划的目的，是在更好地利用景观特色的同时加强对它的保护。

The formation of waterscapes is related to several factors, such as the natural features of site, regional culture, planned usage and target people. The purpose of planning is to make better use of the landscape features and protect them well at the same time.

①自然条件　水作为自然界中循环的物质体，在康复水景的设计中应优先考虑水源的通畅性，保证水的质与量，让水体具有延伸感。

①Natural conditions　As water is circulating in nature, the design of waterscape in a healing garden should give priority to the smooth flow of water sources, then ensure the quality and quantity of water and enable it to be extending and spreading.

水又有形状、形态的自由性。水体具有无限的可变度，是由地形环境及其流向、动力而决定的。了解设计对象本身的地形和水的流速、水量及气候、水温等的特点，预知水的冰、雪、雾等形态的出现是非常重要的。

Water can change freely in shape and form. Therefore, water body has infinite variability determined by terrain environment as well as its flow direction and power. It is very important to understand the features of terrain, the flow rate, volume of water, the climate and water temperature there. Besides, it is key to predict the occurrence of ice, snow, fog and other forms of water.

另外，还需要保证水质及水量的稳定性，水在循环中进行着自然净化，净化也保证了水岸的生态平衡。

In addition, it is necessary to ensure the stability of water quality and quantity. Water is naturally purified in the circulation, purification in turn ensures the ecological balance of water banks.

②水系管理　花园中的水管理应从全园的水系统出发，预测洪水及暴雨对水系统的影响，提出相应的管理对策。康复花园设计中，蓄排水管理是影响水体景观的重要因素。花园中水体的大小是由规划设计的目的、功能需求等因素所决定，但往往在花园建成后，地表径流都会受到影响，所以蓄水方面其容量要按暴雨来进行设计，然后以特定的排量进行排水设计后才能保持园中水体的平衡。具体可以参考雨水花园的设计策略。

②Management of water system　Water management in a garden should start on the basis of the whole water system, which may predict the impact of floods and rainstorms on the system and propose corresponding countermeasures. Water storage and drainage is an important factor affecting the design of waterscape in a healing garden. The size of water body in the garden is determined by several factors, such as the purpose of planning and the functional requirements. In fact, the surface runoff will be affected at the time the garden is completely set up, so in terms of water storage, its capacity should be designed according to the amount of rainstorms, and a plan for specific drainage should be designed to maintain the balance of the water body in the garden. Actually, the water management for healing gardens can refer to the design strategy of rain gardens.

5.3.2.3　边界组织
5.3.2.3　Boundary surroundings

"水随器而成其形"，水没有固定的形状，其形状是由存放它的容器，也就是包围水体的边界组织来决定的。这些组织包括沿水的驳岸、土壤、沙、石、植物、建筑物、构筑物等多种元素。因为有"人"的参与，这些元素也成了与"水"交融的重要媒介。边界组织与水面的开合变化，形成"延而为溪，聚而为池"等不同水体形态。因此，水体与其边界组织的相互关系也就构成了康复水景的不同视觉效果。

A saying goes: the shape of water changes following the different vessels or the containers in which it is stored. In other words, water has no fixed shapes. It is determined by its boundary surroundings including bulkheads along the water, soil and sand, stones, plants, buildings and constructions, etc. Because of the participation of "people", these elements have also become an important medium to interact with "water". The changes of opening and closing between boundary surroundings and water surface may shape water bodies differently, such as the scene "extending to be streams and gathering to be pools". Therefore, the relationship between water body and its boundary surroundings create different visual effects of the waterscapes in a healing garden.

(1)水体及边界组织的构成形式
(1)Water body as well as compositions of its boundary surroundings
水体边界组织的构成与其边界空间的性质、规模、形体、尺度、材料质感及构成方式有关。边界空间具有不同的形态和肌理，并构成特定环境中景观的远景、中景和近景。边界组织的运用方法也有其主要规律，如"掩""隔""破"是古典园林常用的理水形式，可供参考。

The composition of boundary surroundings is related to the nature, the scale, the shape, the criterion, the material texture of boundary space and its ways of composition. The boundary space has different shapes and textures to constitute scenes of three ranges in a specific environment, namely, the long, the medium and the close. Likely, there are also rules in the design of boundary surroundings, such as the form "covering" "separating" and "breaking" that are commonly used for water management in classical gardens.

"掩"是以建筑和绿化，将曲折的池岸加以相互遮掩而又映照衬托的构景方法。如为突出建筑的地位，亭、廊、阁、榭的前部架空挑出水上，水自其下流出，用以打破岸边的视线局限；或者临水以设置蒲苇岸、杂木迷离等方式，体现池水无边的视角印象。"隔"，是筑堤横断于水面，或隔水净廊可渡，或架曲折的石板小桥，或涉水点以步石。如此则可增加景深和空间层次，使水面有幽深之感。"破"即水面很小时，如曲溪绝涧、清泉小池，可用乱石为岸、怪石纵横、犬牙交错，并配置细竹野藤、朱鱼翠藻，如此虽是一洼水池，也令人似有深邃山野风致的审美感觉。

"Covering" is a method of framing a scenery, in which tortuous banks of water body are covered with green belts or hidden behind buildings. All of them reflect each other. For example, to highlight the status of buildings, the fronts of pavilions, corridors, lofts, and terraces are raised above the water, thus the sight limit from the shore is broken; or pampas grasses or weed trees near water can be planted to create a boundless view of lakes or pools. There is also a rule "separating", which is to dam across rivers, build corridor or a small curved stone bridge across the surface of water, or place stepping stones to cross the water. This can increase the sight depth of scenery and the level of space, and create a deeper and quieter scene of water surface. In addition, "breaking" means that when water surface is not so wide, e.g. bending rivers, streams, small springs and pools, many rocks can be set up as a shore with bamboos, wild vines, and some fish and algae will be put in the water. Therefore, it seems to aesthetically create a scenery similar to that of deep mountains and forests, even though it is a small lake or a pool.

（2）亲水心理的运用

（2）Psychology of being intimate to water

最早也许是因为饮水的本能，人总会不自觉地趋向于水边，水体边界是构成人与水之间关系的重要介质。现代快速的城市生活使人们缺少了与自然亲近的机会，于是城市中出现了越来越多的公园，而只要条件允许就会有水景的出现，以满足人们的亲水性需要。康复花园中的水景也正是利用人的亲水心理来达到促进康复的目的。

Because of an instinct of drinking water at first, humans tend to live by the waterside unconsciously. The boundary of water body hence becomes an important medium between people and water. However, rapid urban life at present makes people lack the opportunity to get close to nature, so more and more parks appear in a city. There will be waterscapes, if permitted, to meet people's need to be intimate to water. The waterscapes in a healing garden also rely on such a psychology to achieve the purpose of promoting rehabilitation.

亲水性设计可参考多层次的观赏方式，适当设置亲水平台或进行驳岸的亲水性设计，从而使人们容易亲近水体，也可提供不同尺度的观景感受。

The plan of helping people close to water can refer to the method of multi-level viewing. A platform can be set up or a water bulkhead can be designed to make it easy for people to get close to the water body and provide different views.

（3）确保安全距离
（3）Safety distance

水虽然让人有亲近的感觉，但由于人的生理特性却不能真正地融入水中，适当的距离是必不可少的。人们在康复花园中行走及观赏时，除个别场景需要亲水的功能外，人与水体一般要通过围栏、植物、山石等作为阻断，使人与水保持一定的安全距离。在确保安全的前提下，进行边界组织的无障碍设计，使坐轮椅、挂拐杖等有着特殊需求的群体也能体验亲水的乐趣。

Although water evokes a sense of intimacy, people can't really blend into it because of their physiological characteristics, so a proper distance between them is a must. Generally, when people have a walk and enjoy the scenery in a healing garden, fences, plants, hill-tones, etc. must be set up to block them from water within a certain safe distance, some particular sites where people are allowed to be close to water. Under the premise of safety, ADA designs can be carried out to ensure the groups with special needs, such as those using wheelchairs or crutches, etc. to be able to experience the fun of being intimate and close to water.

自古以来，文人雅士把水誉为高尚品格的象征。老子在《道德经》中写道："上善若水""水利万物而不争，处众人之恶，故几于道。"水滋润万物而不与万物相争，所处位置最不引人注目，所以最接近于道，是为"道法自然"。人、水体与边界之间，应是互为尊重、互为主体、互为彰显、相互促进的关系。总而言之，水体与边界组织是康复花园设计中的重要内容，无论在水体的景观视听方面，还是实际的应用方面，都具有不可忽视的作用和价值。

Since ancient times, Chinese scholars have sung highly of water as a symbol of noble character. Lao Tzu said in his masterpiece *Tao Te Ching*: "Great virtue is like water. Water is a thing that benefits everything yet asks for nothing; a thing that is willing to be ordinary but do anything that others think hard to do, hence it is close to *Tao*, a noble philosophy." Therefore, people and water should respect and promote each other. To conclude, water body as well as its boundary surroundings become key factors in the design of a healing garden, and play an important role in creating the visual and audio effect of waterscape and the practical application.

5.4 造园工程概预算

5.4 Budget of a garden construction project

5.4.1 造园工程概预算的含义

5.4.1 Connotation

造园工程概预算是指在造园工程建设过程中，根据不同设计阶段、设计文件的具体内容和有关定额、指标及取费标准，预先计算和确定园林建设项目的全部工程费用的技术经济文件。

During the process of making a garden, the budget refers to the technical and economic documents in which the total engineering cost of gardening construction will be pre-calculated and determined according to the particular plans at different stages, the related quotas and indicators, and the fee standards.

5.4.2 造园工程概预算的分类

5.4.2 Classification

造园工程建设项目综合性强、涉及面广、环节多，项目从酝酿、提出、决策、设计、施工到竣工验收整个过程中，涉及的各项工作都有先后次序，在工程建设的不同阶段，概预算所做的内容、深浅、粗细都是不同的。

A garden construction project comprehensively involves a wide range of aspects and procedures. From the process of proposing, planning and decision-making, designing, constructing to completion, various procedures involved are sequenced. The budget is different at different stages as to what it is about, and how deep and delicate it will be made.

根据建设活动开展阶段的不同，概预算可分为以下七种。

According to the tasks at different stages of a construction project, the budget can be divided into the following 7 categories.

(1) 投资估算

(1) Budget of investment

投资估算是指在建设项目的决策阶段，对将来进行该项目建设可能要花费的各项费用的事先匡算。

It refers to the various costs that are pre-calculated in the decision-making stage of a construction project.

(2) 设计概算

(2) Budget of design

设计概算是在初步设计或扩大初步设计阶段，由设计单位根据初步设计或扩大初步设计图纸，概算定额、指标，工程量计算规则，材料、设备的预算单价；根据建设主管部门颁发的有关费用定额或取费标准等资料，预先计算工程从筹建至竣工验收交付使用全过程建设费用，编制形成经济文件。简言之，即计算建设项目总费用。

It means that in the preliminary design stage or the expanded preliminary design stage, the designing departments should estimate the quotas, the indicators and the rules of engineering quantity calculation, the unit price of materials and equipments based on the designed drawings at this stage. Moreover, the project administrations should issue relevant documents concerning the cost quota or the expense standard, by which an overall cost of constructing from the beginning to the end can be estimated in advance. In short, a total cost of the project should be calculated.

（3）修正概算

（3）Budget of revision

修正概算是在技术设计阶段，由于设计内容与初步设计的差异，设计单位应对投资进行具体核算，对初步设计概算进行修正而形成的经济文件。其作用与设计概算相同。

The budget of a construction project will be revised during the stage of technical design. Due to the differences between actual situations and preliminary designs, the departments responsible for designing should make specific accounting of investment and revise the preliminary design budget to finally make out corresponding economic documents. The role of this budget is the same as that of the budget of design.

（4）施工图预算

（4）Budget of working drawings

施工图预算是指拟建工程在开工之前，根据已批准并经会审后的施工图纸、施工组织设计、现行工程预算定额、工程量计算规则、材料和设备的预算单价、各项取费标准，预先计算工程建设费用的经济文件。

Before the start of a construction project, constructing, the budget of working drawings should be estimated based on the construction drawings, the design of construction organizations and the current project budget quota, the rules of engineering quantity calculation, the budget unit price of materials and equipment, and various fee standards. All of them have been checked and approved before and finally corresponding economic documents would be made concerning the budget of expense for the project.

（5）施工预算

（5）Budget of construction

施工预算是施工单位内部为控制施工成本而编制的一种预算。它是在施工图预算的控制下，由施工企业根据施工图纸、施工定额并结合施工组织设计，通过工料分析，计算和确定拟建工程所需的工、料、机械台班消耗及其相应费用的技术经济文件。施工预算实质上是施工企业的成本计划文件。

The companies or enterprises responsible for a construction project would make an internal budget in order to control the cost. Under direction of the construction drawing budget, they will calculate and determine the amount of labors, materials and machinery consumption as well as their corresponding expenses that are required by the proposed project, according to the construction drawings, the construction quota and the design of construction organizations. Therefore, the construction budget actually refers to the documents made for planning the project cost.

(6) 工程结算
(6) Project settlement

工程结算是在合同价的基础上调整计算出的结算工程的实际价格。它是发承包双方进行工程价款结算的直接依据。

Project settlement refers to the actual expenses which are adaptably calculated on the basis of the expenses listed in the contract. It is directly a reference for both parties that make the contract to count the whole expenses of the project.

(7) 竣工决算
(7) Final accounts

竣工决算是反映建设项目从筹建起，至建成为止的全部实际成本的技术经济文件，是最终确定的实际工程造价。

The final accounts of a construction project refers to the technical and economic documents that are made to reflect all the projects cost from the beginning to the end. It is a finally actual cost of the project.

5.4.3　工程量清单计价
5.4.3　Valuation with bill quantity(BQ)

5.4.3.1　工程量清单计价的含义
5.4.3.1　Connotation of BQ pricing

根据住房和城乡建设部《建设工程工程量清单计价规范》(GB 50500—2013)的有关规定，部分使用国有资金投资的建设工程施工发承包，必须采用工程量清单计价；非国有资金投资的建设工程，宜采用工程量清单计价。

According to *Code of Valuation with bill quantity of Construction Works* (GB 50500—2013) issued by the Ministry of Housing and Urban-Rural Development, the construction projects that are partially invested by state-owned capitals (hereinafter refers to as state-owned investments) must use BQ for pricing. So do those projects invested by non-state-owned funds.

工程量清单计价的造价组成应包括按招标文件规定，完成工程量清单所列项目的全部费用，具体包括分部分项工程费、措施项目费、其他项目费和规费、税金。工程量清单应采用综合单价计价，即综合单价中包括完成一个规定清单项目所需的人工费、材料费、施工机械使用费、企业管理费、利润，以及一定范围内的风险费用。即综合单价包括除规费、税金以外的全部费用。综合单价不但适用于分部分项工程量清单，也适用于措施项目清单、其他项目清单等。

In accordance with corresponding bidding documents, BQ pricing should include all the costs of completing the items listed on the bill: sub-item engineering fees, project measure fees and other project fees, regulatory fees and taxes. The pricing adopts the method of comprehensive unit price. It means that the unit price should include the costs of labors and materials, the fees for machinery usage and management, and profits required to complete those qualified products of each measurement unit. In addition, the costs of risks within a certain range should be taken into consideration. That is to say, the comprehensive unit price includes all expenses except regular fees and taxes. The comprehensive unit price is not only applicable to sub-item BQ, but also to the pricing list of measures project and other items.

5.4.3.2　工程量清单计价步骤
5.4.3.2　Steps of BQ pricing

①研究招标文件，熟悉工程量清单、施工图纸及相关资料。

① Study the bidding documents; get familiar with corresponding BQ, construction drawings and related materials.

②考察现场，了解地形、地下情况及周边环境等。

②Inspect the site to know well its terrain, underground conditions, its surroundings and the like.

③了解施工组织设计，熟悉加工订货的有关情况，明确主材和设备的来源情况，确定施工方案。

③Understand the design of constructing organizations; get familiar with the procedures of processing and ordering goods; clarify the source of main materials and equipment, and work out the construction plans.

④计算工程量，包括两部分内容：一是核算工程量清单所提供清单项目工程量是否准确；二是计算每一个清单主体项目所组合的辅助项目工程量，以便分析综合单价。

④Calculate project quantity. This includes two parts: one is to calculate whether the quantity of project items on the bill is accurate or not; the other is to calculate the quantity of those projects auxiliary to each main project on the bill, so as to analyze the comprehensive unit price.

⑤确定措施项目清单内容，措施项目清单是完成项目施工必须采取的措施所需的工作内容。因此，在确定措施项目清单内容时，必须结合项目的施工方案或施工组织设计的具体情况填写，必要时要对施工方案或施工组织设计加以修改。

⑤Determine the list of measure items. The measure item list refers to the work content required to take measures to complete the construction of a project. Therefore, when the list of measures project is made, it must be filled out in accordance with the work plan or the process design of the construction project, and it must be, if necessary, modified according to real situations.

⑥选定参照定额，确定管理费和利润，确定人、材、机单价。

⑥Select the reference quota, determine the management fee and the profit, and decide the unit payment for personnel, and the unit price for materials and machines.

⑦分析清单项目的综合单价。

⑦Analyze the comprehensive unit price of the items on the list.

⑧计算措施项目费、其他项目费、规费、税金等，将分部分项工程项目费、措施项目费、其他项目费和规费、税金汇总、合并、计算出工程造价。

⑧Calculate measure item fees, other item fees, stipulated fees, taxes, etc.; count the whole project cost by accumulating the fees of sub-item projects, measure items, other items, stipulated fees and taxes.

⑨按招标文件的要求，填写表格，并装订成册。

⑨Fill in relative forms as required in the bidding documents and bind them into a volume.

5.4.3.3　工程量清单计价包含的费用
5.4.3.3　Expenses included in BQ pricing

工程量清单计价价款，应包括完成招标文件规定的工程量清单项目所需的全部费用。其内容如下：

A BQ pricing should include all the expenses required to complete the BQ items specified in bidding documents. These expenses include：

①包括分部分项工程费、措施项目费、其他项目费、规费和税金；

①The fees of sub-item project, measure items and other items; stipulated fees and taxes.

②包括完成每分项工程所含全部工程内容的费用；

②The cost of completing each sub-project.

③包括完成每项工程内容所需的全部费用(规费、税金除外)；

③The expenses required to complete each project (except stipulated fees and taxes).

④工程量清单项目中没有体现的，施工中又必须发生的工程内容所需的费用；

④The expenses required for those work that must be completed in constructing process yet not included in the items of BQ.

⑤考虑风险因素而增加的费用。

⑤Increase of expenses in consideration of risks.

作为新时代的园林人，需要掌握园林项目工程投资的构成、各分项工程成本计算及控制，了解具体工程概预算的方法及文件编制，建立现代科学工程造价管理的思维观念和方法，具有工程造价管理的初步能力。

As gardeners in a new era, they need to understand the compositions of a garden project investment, the cost calculation and the control of each sub-project. Meanwhile, they need to know the methods of making specific project budgets as well as corresponding document preparation, establish scientific modern concepts and methods of engineering cost management, and have the primary capability of managing project cost.

由于篇幅关系，更多关于造园工程概预算的内容可以参考相关书籍。

For more information about the budget of garden projects, please refer to other books.

思考题

1. 简述康复花园的设计目标与设计原则。
2. 康复花园与普通花园在设计要素方面有什么区别？
3. 你对康复花园药用花卉植物造景有哪些看法？

Questions

1. Please briefly describe the design goals and principles of healing gardens.
2. What is the difference between healing gardens and ordinary gardens in terms of the design elements?
3. What are your views on the landscaping of medicinal flowers and plants in healing gardens?

第6章

医院康复花园设计特点

Chapter 6　Design Features of Hospital Healing Gardens

　　我国现有医疗机构尚未把康复花园列为外部环境设计的基本元素之一。据北京、上海、广州、深圳等几大城市医疗机构外部空间现状调查显示：大多数医疗机构人满为患、拥挤不堪，患者挂号、就医等环节都较为紧张，因此不少医院外部空间多被停车场或其他医疗设施占据，绿地面积少，且缺少休息设施。可见在医院绿地率无法得到保障的前提下，康复花园建设显然难以提上议程。

　　The existing medical institutions in China have not listed the healing garden as one of the basic elements of the external environment design of medical institutions. A survey on the current status of the external space of medical institutions in major cities such as Beijing, Shanghai, Guangzhou, and Shenzhen has shown that most of those institutions are overcrowded, thus straining all procedures such as making an appointment and seeking medical advice. Most of the external spaces of hospitals are occupied by parking lots or other medical facilities, so the green is small and the rest facilities are deficient. The coverage of green space in most hospitals cannot be guaranteed, let alone the construction of "healing gardens".

　　为改善我国医疗机构的绿色环境空间品质，为患者、医护工作者、来访者等人群提供舒适优美的就医环境，笔者希望能够借由此书及诸多有识之士的共同努力，逐步引起社会各界对现有医院绿色环境建设的重视。

　　In order to gradually improve the quality of the green environment in China's medical institutions and provide a comfortable and beautiful environment for patients, medical staff, visitors, etc, the author hopes that this book, together with the joint efforts of many enlightened individuals, can gradually raise awareness among all sectors of society about the importance of constructing green environments in existing hospitals.

6.1　使用人群
6.1　Users

医院环境的主要使用者包括以下三类群体：患者、医护人员、探访者。其中，患者是医院服务的主体，根据患者的疾病类型，可以分为多种群体。患者由于身体或者心理上的疾病较正常人更容易出现心理紧张、烦躁、焦虑等负面情绪，更容易受到外界环境的影响。近年来部分研究结果表明：优美的室内外环境能够帮助患者减轻疾病带来的痛苦，提升愉悦感，同时也能在一定程度上加快患者的康复。

The main users of the hospital environment include the following three groups: patients, medical staff, and visitors. Among them, patients are the main body served by the hospital, and can be divided into multiple groups according to their disease types. Due to physical or psychological diseases, patients are more likely to have negative emotions such as nervousness, irritability, and anxiety than normal people, and to be affected by the external environment. In recent years, some researches have shown that a beautiful indoor and outdoor environment can help relieve the patients of the pain caused by diseases and enhance their pleasure, and can also speed up their recovery to a certain extent.

6.1.1　患者
6.1.1　Patients

患者能否使用优美的户外环境很大程度上取决于他们的健康状况，健康状况决定了他们外出的能力。那些没有携带监测设备，或者不需护士的帮助自己就能活动的，以及长期护理的患者，是最乐意使用户外空间的群体，具体包括：整形外科患者，瘸腿或者其他类型的骨折患者，产科护理患者，康复中的患者，精神病患者，等等。总之，这些处于恢复期的患者均可独立活动。其他重症患者，如癌症、肺结核、烧伤、术后患者等，他们到户外去的能力，则取决于医护人员或探访者的帮助。

Whether patients can take advantage of a beautiful outdoor environment largely depends on their health status, which determines their possibility to go out. Those who don't need to carry monitoring devices, or can move around without the help of a nurse, and who require long-term care are the groups the most willing to use the outdoor space. They include: patients undergoing plastic surgery, patients with a broken leg or other types of fractures, patients requiring obstetric nursing, patients in recovery, patients suffering from mental illness; in short, those types of patients can move independently to some extent during their recovery. For other critically ill patients, such as postoperative patients, those suffering from cancer, tuberculosis, and burns, the ability to go outdoor depends on the help of hospital staff or visitors.

另外，因为疾病会降低肌体适应外界环境的能力，患者在户外温度和风的作用下更容易加重病情，即患者对外界环境条件的反应比健康人更为敏感。例如，烧伤患者对热和阳光很敏感。所以，患者需要在使用户外空间时拥有选择权。

Diseases reduce human body's ability to adapt to outside temperatures. Therefore, their diseases are more likely to get worse due to outdoor temperatures and wind, and they are more sensitive to the environment than healthy people. For example, burn patients are sensitive to heat and sunlight. Therefore, patients need to have the right to choose when using outdoor space.

6.1.2　医护人员
6.1.2　Medical professionals

医护人员是医院中为患者提供服务的主要群体，医生负责疾病的诊治，而护士负责患者的护理。近年来由于各种疾病高发、各类医院人满为患，导致医护工作者日常工作压力很大，越是在这样的条件下，越需要提升医院的环境品质，为医护工作者提供减压放松的场所，帮助医护工作者调节工作压力，从而更好地提高服务质量。

Medical professionals are the main group that provide services for patients in the hospital. Among them, doctors are responsible for the diagnosis and treatment of diseases, while nurses are responsible for patient care. In recent years, due to the high incidence of various diseases, all types of hospitals are overcrowded, and the medical staff are under great daily pressure. It is such a situation that makes it more necessary to improve the quality of hospital environment. The services would be improved by providing some places to help medical staff take pressure off their work and revitalize themselves.

6.1.3　探访者
6.1.3　Visitors

探访者主要指探望或者陪护患者的亲朋好友。探访者是患者的支持者，让患者保持着与外界的联系，同时，探访者对患者的关心和照顾有利于患者排解不良情绪，早日康复。因此，医院需要为探访者提供一个能够与患者放松交流的空间。

Visitors mainly refer to the relatives and friends who visit or accompany the patients. They are usually the patients' supporters, enabling them to maintain contact with the outside world. Meanwhile, the visitors' care and concern for patients is conducive to their elimination of bad emotions and speedy recovery. Therefore, hospitals need to provide visitors with a place to relax and communicate with patients.

上述每一类群体都有其特点和需求，医院户外空间设计要充分考虑各群体的特点，有针对性地进行设计。

Each of the above-mentioned groups has its characteristics and needs. The characteristics of each group should be fully considered in the targeted design of outdoor spaces of the hospital.

6.2 设计程序与设计要点
6.2 Design procedures and key points

6.2.1 选址
6.2.1 The site selection

医院等医疗机构室外环境设计最好从医院的选址开始，因为一座位于城郊、环境优美、用地充足的医院，能够为建立良好的康复花园和医院环境提供最佳的用地条件和保证，如美国宾夕法尼亚州 State college 郊外 Mount Nittany Medical Center（图 6-1）；相反，如果医院选址于闹市，不仅用地范围受限，而且周边环境也相对嘈杂，很难提供丰富的景观资源。现在位于大城市中心的医院往往如此。因此，建议负责医院环境设计的景观设计师，能够参与新建医院前期的选址和用地大小的决策过程。

For the outdoor environment design of medical institutions, it is best to start with the site selection of the hospital, for instance, a hospital located in the suburbs with a beautiful environment and sufficient land can surely provide the best land conditions and guarantees for the establishment of a good healing garden and hospital environment, for example, Mount Nittany Medical Center (Fig. 6-1); on the contrary, if the hospital is located in a downtown area, not only is the land for use limited, but the surrounding environment is relatively noisy, which makes it difficult to provide abundant landscape resources. This is often the case for hospitals located in the center of large cities. Therefore, it is recommended that the landscape architects responsible for the hospital's environmental design participate in the decision-making process of the site selection and land size of the new hospital in the early stage.

图 6-1 **Mount Nittany Medical Center** 周边环境（吴祥艳 摄）
Fig. 6-1 The site selection of Mount Nittany Medical Center (by Wu Xiangyan)

在我国，医院或者医疗卫生机构作为医疗卫生服务设施用地，是城市用地的重要组成部分，其规划数量、规模和位置需要综合考虑城市人口规模、人口分布、服务半径、疾病谱等因素。遵从《全国医疗卫生服务体系规划纲要》等相关规范，医院的选址和用地大小往往取决于一个城市或城镇的总体发展规划。建议景观设计师参与到总体规划编制过程中，协助规划师选择最优的地段作为新建医院的基址。

In China, as medical and health service facilities, hospitals or medical and health institutions are an important part of the urban land. Factors such as the urban population size, population distribution, service radius, disease spectrum need to be considered for the planned quantity, scale and location of hospitals or medical and health institutions. In compliance with the *Outline for the Planning of the National Medical and Health Service System* and other relevant regulations, the location and land size of a hospital often depend on an overall plan for the development of a city or town. It is recommended that landscape architects participate in the process of drawing up an overall plan and assist planners in choosing the best site for the construction of a new hospital.

6.2.2　对户外空间的总体要求（总体规划阶段）
6.2.2　**Overall requirements for outdoor space**（at the stage of master planning）

户外空间的使用方式，部分取决于临近建筑或空间的使用，以及潜在使用者。克莱尔·库珀·马库斯在《人性场所——城市开放空间设计导则》一书中建议在医院总体规划之始，就邀请景观设计师参与，协助确定户外空间的位置、方向以及可达性。"要将整个场地作为一个整体的治疗环境考虑，景观设计师作为跨学科设计团队（IDT）的成员必须在规划设计最初阶段加入，要邀请参加过康复景观设计训练的景观设计师参与"（克莱尔，2001），除此以外，还需要考虑以下方面。

The way the outdoor space is utilized depends in part on the use of nearby buildings or space as well as potential users. Claire Cooper Marcus suggested in the book *People Places: Design Guidelines for Urban Open Space* that at the beginning of a hospital's master plan, landscape architects should be invited to join the team to help determine the location, direction and accessibility of outdoor space. "The whole site must be considered as a healing environment; the landscape architects, as members of the interdisciplinary design team (IDT), must participate in the initial stage of planning and design; and landscape designers who have undergone the healing landscape design training must be also invited to participate" (Kelaier, 2001). In addition, the following aspects also need to be considered.

①建筑设计和花园选址同时考虑。

①The architectural design and garden site selection should be considered at the same time.

建筑风格、各功能空间与花园的位置关系，从室外望向花园的视野等，均决定花园的使用效率。因此，在总体规划阶段，就要考虑将主花园设置在公众主要使用的空间，如大厅、等候区、走廊或者餐厅能够看到的地方，或者容易到达的位置。

The efficient service is determined by the architecture style, the positioning between each function space and the garden, and the view of the garden from the outside, etc. Therefore, in the stage of overall planning, it is necessary to consider locating the main garden in the position, which is easily seen or accessible from such public space as lobbies, waiting areas, corridors or canteens.

②提供多样的户外空间。

②Diversified outdoor spaces should be provided.

因为有多种不同的使用者(短期住院患者、长期住院患者、员工、探访者)，所以应提供多样化的户外空间。

As there may be different groups of users (short-term and long-term inpatients, staff, visitors), diversified outdoor space should be available.

③最少提供一处可以让使用者感觉远离身外世界或者医院环境的空间。

③At least one place that allows users to stay away from the outside world or the hospital environment should be provided.

大多数人都想寻找一处安静的隔离环境，那里拥有满眼绿意、鸟语花香、新鲜空气等可以激发人们感觉的元素。

Most people want to seek a peaceful and isolated environment, where birds' twitter and fragrance of flowers, an eyeful of greenery and the fresh air can refresh people.

④把需要鼓励使用户外空间的患者与员工安排在离花园入口最近的位置。

④The patients and employees who need to be encouraged to use outdoor space should be arranged at the location closest to the garden entrance.

相反，有些患者禁止使用户外空间，如呼吸道疾病患者，则应该安置在最远离花园入口的房间中。

In contrast, some patients who are forbidden to go outdoors, such as those with respiratory diseases, should be placed in the room furthest away from the garden entrance.

⑤在花园两侧有窗户相对的地方，花园的宽度不小于6m。这个距离给所有人提供了足够的私密性，特别是透过窗户以及窗外的花园可以看到对面窗内的病房或者办公室，这个距离更为重要。

⑤The width of the garden shall be at least 20 feet (about 6 meters) where the window on both sides are directly opposite. This distance provides sufficient privacy for all those involved. Especially when the wards or offices in the opposite can be seen through the windows and the outside garden.

这个距离给所有涉及者都提供了足够的私密性，特别是在有窗的走廊中，如果透过户外空间可以看到病房或者员工办公室，这一点更为必要。

This distance especially in corridors with windows, provides enough privacy for all people involved. It is even more important if patient wards or staff offices can be seen from the outdoor space.

⑥户外花园，尤其是主要花园，无论采取哪种形式，均要仔细分析小气候条件，并在设计时保留足够的光照空间。

⑥Despite a variety of forms, the microclimate conditions must be carefully analyzed, especially for major gardens. Enough spaces with natural light should be reserved in the planning of outdoor gardens.

光照充足能够提升北方地区户外空间的使用率，人们可在清凉的早春、深秋季节，甚至暖和的冬日更多地使用户外空间。反之，夏季人们则需要更多的荫蔽场所来躲避强烈的阳光。对于主花园，至少保证一天接近 6 小时的日照。同时，也要考虑风的因素，避免冬季冷风直吹，夏季则要有凉风可以进来。

Adequate sunlight allows people in the northern regions to use the outdoor space more in the cool early spring and late autumn, as well as in the warm winter days. In contrast, people in summer need more shaded areas to avoid strong sunlight. For the major gardens, at least nearly six hours of sunshine a day should be guaranteed. Apart from that, the setting of the main gardens should take into account the factor of wind, avoiding the direct exposure to the cold wind in winter, and at the same time, there should be cool breeze in summer.

⑦如果花园四周没有建筑物围合，则应该用栅栏或者树篱充分封闭。

⑦If a healing garden is not completely surrounded by buildings, it should be fully enclosed with fences or hedges.

当花园紧靠道路或者停车场时，让使用花园的人不受花园外部事物干扰，保证充分的安全和私密性。这一点尤其重要。

This is especially important when the garden is close to a road or parking lot. People who use the garden should feel safe and private without being disturbed by things outside the garden.

⑧避免将户外空间安置在紧邻散热口或空调设备的地方。散热口和空调设备通常有很大的噪声、气味或不适宜的温度等，因此，要避免户外空间的使用者受到这些因素的干扰。

⑧The outdoor spaces should not be placed near heat pumps or air-conditioning equipments, which usually produce noises and odors or lead to unsuitable temperatures, etc. Therefore, it is necessary to prevent users of outdoor spaces from being disturbed.

⑨为医院员工提供一处不轻易被患者或探访者看到的户外空间。

⑨Hospital staff should be provided with an outdoor place that is not easily visible to patients and visitors.

目的在于为员工提供一个能够暂时摆脱处理患者的压力与紧张状态的空间，缓解医护人员的压力。

The purpose is to provide the staff with a place where they can be temporarily free from the pressure and tension of dealing with patients.

⑩设置清晰的通往花园或者花园内部的标识系统。

⑩It's necessary to provide clear signs leading to the garden or its inside.

当花园非一眼能见，如屋顶花园或者中庭花园，须设置清晰的导向标识。

When the gardens such as a roof garden or an atrium garden are not visible at first sight, it is very important to set up clear signs.

⑪设计要考虑维护一种包容的文化，服务所有使用医疗服务的人。

⑪It's necessary to maintain an inclusive culture to serve all people who use medical services.

⑫增设兼具休息和餐饮的室内候诊空间。

⑫It's necessary to build the indoor waiting space that can be used for both rest and dining.

在欧美地区，医院内部常设计配置小型的餐饮区，方便前来就诊的患者和家属就餐。同时，就餐区域附近设置一处景色优美的花园，方便人们休憩放松。在信息科技快速发展的今天，伴随着线上诊疗的发展，位于大城市中心区域的医院，一方面可以借助线上诊疗缓解医院实体空间的不足；另一方面，可以尝试将一些使用效率不高的室内空间或者建筑屋顶等，改造成具有休息、放松、餐饮等功能的室内花园或屋顶花园，结合线上小程序，为患者提供 30 分钟就诊提醒，让患者和家属无须停留在拥挤的病房外候诊。把花园搬到室内，可以在北方冬季提高其使用效率，以改善患者的候诊体验。或者在建筑设计时将毗邻病房的候诊区域加大，增加一定的景观元素和简餐饮料等，提升候诊区域的舒适度。

Many European and American hospitals often set up a catering area for patients and their families. At the same time, a beautiful garden is set up near the dining area for people to have a rest and relax. Today, information technology has developed rapidly. Therefore, with the help of online diagnosis and treatment, the hospitals located in the central area of big cities can, on the one hand, alleviate the shortage of hospital physical space. On the other hand, they can try to transform some inefficient indoor spaces or building roofs into indoor gardens or roof gardens working. They can provide people with rest, relaxation, catering, etc., and also provide patients with 30-minute medical reminders in combination with online reporting app., so that the patients and their family members do not need to stay outside crowded wards while waiting to see the doctor. Moving the garden indoors can increase its usage efficiency in northern winters and improve patients' waiting experiences. Or the waiting area adjacent to the ward will be enlarged in the architectural design, and certain landscape elements, light meals and soft drinks will be added to improve the comfort of the waiting area.

6.2.3　医院康复花园的位置、类型与特点
6.2.3　The location, types and characteristics of healing gardens in the hospitals

上文强调：首先需要将整个医疗环境看作是一个完整的"康复花园"，将建筑与景观一体化考虑。那么，就室外空间而言，不同位置的康复花园该如何设计？其优点和缺点如何？鉴于国内尚无此方面较为系统的研究成果，笔者对欧美国家康复性景观的相关研究进行梳理汇总，希望能够对我国医疗机构户外空间设计和革新提供参考，

对学生的学习有一定启发。

It has been emphasized in the preceding part of the textbook: first, the entire medical environment needs to be regarded as a complete "healing garden" integrating architectures and landscapes. Then, from the perspective of outdoor space, how should the healing landscapes be designed in different locations? What are its advantages and disadvantages? Since there is no systematic research in this field yet in China, the author collects and summarizes the related researches on the healing landscapes in European and American countries, in the hope of offering a reference for the outdoor space design and innovation of medical institutions in China, and of providing inspiration for students.

在欧美地区，依据空间位置的不同，将医疗服务设施中的康复性景观大致划分为以下类型：前廊、入口花园、后院花园、隐秘花园、庭院、广场、屋顶花园、屋顶露台、中庭花园、纯观赏花园等，各个类型花园的特点和设计现状总结如下。

In European and American countries, the healing landscapes in medical service facilities are roughly divided as follows according to different spatial locations: Front Porch, Entry Garden, Backyard Garden, Tucked-Away Garden, Courtyard, Plaza, Roof Garden, Roof Terrace, Atrium Garden, Viewing Garden, etc. The characteristics of each type are summarized as follows.

6.2.3.1　前廊
6.2.3.1　Front porch

在医疗机构主入口处，通常会设计一处以硬质铺装为主的区域，类似住宅的前廊（图6-2）。这一区域是患者进入医院的第一站，主要承担步行和车行出入的功能，满足患者上下车、进出医院的安全。如果空间较大，可以设置座椅、导诊标识、邮箱、电话等设施。前廊空间是否舒适对患者的心情非常重要：是令人放松的、安静的、吸引人的？还是令人迷惑的、拥挤的、不安的呢？

The main entrance of a medical institution is usually designed with a hard pavement area, namely, a front porch, similar to that of a house (Fig. 6-2). This area is the first stop for patients to enter the hospital, and therefore mainly for people to walk or for cars to go in and out, or for patients to get in and out of the car. If the space is large, some facilities, e. g. seats, guide signs, mailboxes, telephones can be set up. It is very important to patients whether the front porch space is comfortable or not: is it relaxing, quiet and attractive? Or is it confusing, crowded and uneasy?

（1）理想设计
（1）Ideal design

① 提供与主入口的视觉联系，类似住宅的前廊；

①The front porch similar to that of a house, provides visual cues to the main entrance;

② 悬垂的前廊可以缩小建筑的体量；

②The overhanging front porch may reduce the volume of building;

图 6-2 Mount Nittany Medical Center 门诊入口门廊（吴祥艳　摄）

Fig. 6-2 The front porch of Mount Nittany Medical Center（by Wu Xiangyan）

③精心设置的休息区可以形成标志性前廊景观，让那些等待被接送、等候公共汽车或其他过往者感到舒心。

③The well-designed rest area can become a distinctive front porch landscape，making those waiting to be picked up，waiting for buses，or just looking around feel comfortable.

（2）非理想设计

（2）Non-ideal design

①如果是通过建筑物地下的停车场或者临近的停车场才能进入该机构的主入口，则可能导致前廊空间不足，难以满足功能需求；

①If people get to the main entrance to the institution through the parking lot beneath the building or through the adjacent parking lot，the front porch may be underused；

②如果车行和人行出入没有很好地整合，或者医院的主入口和急救出口毗邻，可能使人无所适从；

②If the needs of vehicles and pedestrians are not well integrated，or when the main entrance of the hospital is adjacent to the emergency exit，the front porch may make people confused；

③如果有吸烟者在此停留，容易引起行人不快。

③If there are smokers lingering here，it can easily be an unpleasant experience for passersby.

6.2.3.2　入口花园

6.2.3.2　Entry garden

入口花园是靠近建筑主入口的景观区域（图 6-3）。与前廊不同，入口花园是一处具有花园形象的绿色空间，是经过设计且用途明确的空间。

图 6-3　门诊入口边花园(吴祥艳　摄)

Fig. 6-3　The entry park (by Wu Xiangyan)

An entry garden is a landscape area close to the main entrance to a building (Fig. 6-3). Unlike the front porch, it is a green space with a garden image and is designed with a clear purpose.

(1)理想设计

(1) Ideal design

①入口花园易见，且可以从常用入口进入；

①The entry garden is easy to see and people can enter it from the usual entrance；

②尽可能美化入口区域，否则可能会被误作停车场；

②The entry garden should be beautified as much as possible, or it may be mistaken as a parking；

③为使用者提供一个令人愉快的"花园"形象；

③It can provide a pleasant "garden" image for its users；

④允许门诊患者或附近老年居民使用；

④Ambulatory inpatients or elderly residents nearby are allowed to use it；

⑤可拓展为公园，从市政机构获得养护费用。

⑤It can be successfully used as a public park and receive maintenance fees from municipal government departments.

(2)非理想设计

(2) Non-ideal design

①花园的尺寸和位置可能降低主入口的可达性；

①The size and location of the entry garden may reduce the accessibility to the main entrance；

②缺少私密空间可能会阻碍住院患者使用。

②The lack of privacy may hinder inpatients to come.

6.2.3.3　后院花园
6.2.3.3　Backyard garden

后院花园专指位于建筑后面的花园。

A backyard garden refers specifically to the garden behind a building.

(1)理想设计

(1)Ideal design

①后院可以提供一个安静的绿色空间，与前廊和入口花园的繁忙形成对比；

①The backyard can provide quiet green space, in contrast to the busy front porch and entry garden;

②如果后院设计得当，会让那些住养老院或临终关怀医院的人看到花园时，愉快地想起自家后院，有很强的归属感。

②If the backyard is planned properly, those who live in nursing homes or hospices will happily think of the backyard of their home when seeing the garden.

(2)非理想设计

(2)Non-ideal design

①如果没有清晰的标识，偶尔到访该机构的人可能不知道后院花园的存在；

①People who visit the medical institution occasionally may not know the existence of the garden without clear signs;

②如果邻近停车场或街道等，可能无法满足人们对安静环境的要求；

②If there are parking lots or streets nearby, people's requirements for a quiet environment may not be met;

③如果后院花园是为失智症患者或精神科住院患者而设的，须仔细设计及伪装花园的边界，以防止患者逃离。

③If it is designed for patients with dementia or inpatients of the psychiatry department, the boundary of the garden must be carefully designed and camouflaged to prevent them from escaping.

6.2.3.4　隐秘花园
6.2.3.4　Tucked away garden

隐秘花园指远离建筑或者被临近道路、停车区域或者服务入口分隔开的花园。

A "tucked away" garden refers to a garden far from buildings or separated by adjacent roads, parking areas or service entrances.

(1)理想设计

(1)Ideal design

①能充分利用"闲置"空间；

①The "tucked away" garden can make full use of the space that may be left unused before;

②走一小段路到花园可获取一个愉快的、逃离建筑内部活动的感受。

②People can get a very pleasant feeling of escaping from what they have to do in the building only by walking a short distance to the garden.

（2）非理想设计

（2）Non-ideal design

①没有良好的标识或可见性，则使用率会降低；

①Without a good signage or visibility, the usage rate of the garden will be very low;

②在养老机构中，如果可达性不高则影响使用。

②In senior facilities, it is unlikely to be used fully if it is difficult to access.

6.2.3.5　庭院

6.2.3.5　Courtyard

庭院空间通常为正方形、长方形，或者"L"形，四周通常是建筑墙体和窗户围合。庭院是建筑群组中的虚空间，便于采光。医院的庭院最好显而易见，或者是一进大门就可以看见，让患者或访客快速了解花园的存在。

A courtyard is usually square, rectangular, or L-shaped, surrounded by building walls and windows. It is a virtual space in the building group, which is convenient for lighting. It is best for the hospital courtyard to be clearly visible, or to be seen as soon as patients or visitors enter the gate so that they can know the existence of the garden immediately.

（1）理想设计

（1）Ideal design

庭院被建筑围合，呈现半私密的特点，为住院患者和特殊患者提供高度的安全性（例如，成为精神病患者使用的单元或阿兹海默病患者的设施）。

The inner courtyard is enclosed by buildings and presents a semi-private feature, providing inpatients and special patients with a high degree of safety (for example, it becomes units used by mentally ill patients or the facilities for patients with Alzheimer disease).

①虽然庭院自身的光照会受到周边建筑高度的影响，但可以为建筑物的中心带来采光；

①Bring light to the center of the building, though the lighting of the courtyard itself will be affected by the height of the surrounding buildings;

②相邻建筑物可以保护空间免受风吹或提供必要的阴凉，如果建筑物不太高的话，空间会更符合人的尺度；

②Adjacent buildings can protect the space from wind or provide necessary shade. If the building is not too high, the scale of the space would be more people-friendly;

③当室内自助餐厅占据庭院的一个或者多个边角时，可以兼作户外就餐区使用；

③When the indoor buffet restaurant occupies one or more corners of the courtyard，it can also be used as an outdoor dining area；

④庭院可以为周围办公室或者病房提供优美的窗景；

④The courtyard can provide beautiful window views for the surrounding offices or wards；

⑤庭院内特色植物造景可以引导使用者前行。

⑤The characteristic plant landscape in the courtyard can guide users to move forward.

（2）非理想设计
（2）Non-ideal design

①庭院使用效率高低取决于其大小、位置和设计。有些庭院可达性较高，也有些则远离入口和主要视线，降低使用效率。

①The efficiency of courtyard depends on its size，location and design. Some courtyards have high accessibility，while others are far away from the entrance and main sight.

②庭院的"鱼缸效应"会导致很少有人愿意使用这个空间，因为使用者会感觉被监视。

②The "fish tank effect" of the courtyard will make few people willing to use this space because they feel "displayed"，namely，they think their privacy is violated.

③如果缺乏足够的绿化或者结构上的缓冲区，会导致临近庭院的房间和庭院存在不可避免的视线和声音的相互干扰。

③If there is insufficient greenery or structural buffer zone，it will inevitably lead to the interference of sight and sound between the adjacent rooms and the courtyard.

④如果是出于自然采光或交叉通风需要而设置的庭院，若无足够的预算保证庭院的良好状态，庭院的使用可能会受到限制。

④If the courtyard is set for natural lighting or cross ventilation，but there is no enough budget to keep it in good condition，and the use of the courtyard may be restricted.

⑤大多数庭院太小，甚至设有嘈杂的暖通空调机组设备，不能提供足够的活动空间。

⑤Most courtyards are too small，even equipped with noisy HVAC units，to provide enough space for activities.

6.2.3.6　广场
6.2.3.6　Plaza

医院的广场空间主要是那些硬质表面的空间，可以有树木、灌木和花卉等的种植池，但不以绿地为主，而是以硬质场地为主。

Plaza spaces of the hospital are mainly those with hard surfaces，and there can be planting beds of trees，shrubs and flowers，but a plaza mainly consists of a hard landscape instead of a soft one.

（1）理想设计

（1）**Ideal design**

①植物养护和灌溉费用低廉；

①The cost of plant maintenance and irrigation is low；

②场地不大，如果设计合理，也会吸引游人；

②Regardless of limited venue, it will attract a lot of people if the design is reasonable；

③关注铺装细节，确保轮椅患者、步行、挂拐杖者使用方便；

③Attaching more importance to the details of the pavement, make sure that it is convenient for patients using a wheelchair, walkers and crutch users；

④种植大型乔木形成绿荫，并在绿荫下安排各种休息性座椅，能够形成良好的坐息空间或集散空间。

④Large trees are planted to form the green shade, under which various seats are arranged to form a good place for resting and gathering.

（2）非理想设计

（2）**Non-ideal design**

①广场通常缺少人们所认为的康复空间的特征，如绿意盎然、五彩缤纷；

①The plaza usually lacks the characteristics that people think a healing place should have, such as greenery and rich colors；

②广场人员嘈杂的时候（国内很多医院广场如此）会让人联想到购物广场或者商厦，而不是一个平静的、减压的区域；

②A hubbub of voices（as in many hospital plazas in China）will make people reminiscent of shopping plazas or commercial buildings, rather than a calm and pressure-relieving area；

③夏季硬质表面由于反射光线，温度会很高，影响使用；

③The heat reflected from the hard surface of the pavement in summer may make the plaza unsuitable for use；

④浅色的铺装容易刺眼，尤其是在阳光充足的夏季，对老年人和使用某些特殊药物治疗的患者不利；

④The light-colored paving is prone to be dazzling, especially in the sunny summer, which is harmful to the elderly and the patients treated with special drugs；

⑤如果铺装的伸缩缝大于0.3cm，患者使用的助行架或静脉输液架的轮子可能会陷入其中，造成不便；

⑤If the expansion joint of the pavement is larger than 0.3cm, the patients who use a walking frame or IV pole might be affected, because the wheels may get stuck；

⑥纯粹的硬质广场会有冰冷坚硬之感，让人不愿亲近。

⑥A purely hard-landscaping plaza makes people feel cold, hard and reluctant to approach.

国内医疗机构中面积较大的广场一般是门诊楼入口广场、停车场等。这些广场的设计在满足基本功能的前提下，尽可能增加绿地、树木以及座椅等休息设施，以丰富广场空间景观并提供更舒适的使用功能。停车场等硬质广场也需要设计成花园式。

The plazas with a relatively large area in domestic medical institutions are generally those at the entrance of the outpatient building, the parking lot, etc. The design of the plaza in a healing garden, on the premise of playing its essential role, should increase the rest facilities such as green area, trees as well as seats as much as possible to enrich the space landscape of the plaza and make people feel more comfortable. Hard-landscaping plazas such as parking lots also need to be designed in a garden style.

6.2.3.7　屋顶花园
6.2.3.7　The roof garden

屋顶花园指在医院的屋顶上建设的供患者、医护人员或者探访者使用的花园，也有一些是供办公区或者病房区眺望的。与庭院相比，屋顶花园可以向多个方向眺望（图6-4）。

A roof garden refers to the garden built on the roof of the hospital for patients, staff or visitors to enjoy. Some can be used to look into the distance from above for people in the offices or wards. Compared with the courtyard, people can look into the distance from the roof gardens in different directions (Fig. 6-4).

图 6-4　远眺(吴祥艳　摄)

Fig. 6-4　Looking into the distance from the roof gardens (by Wu Xiangyan)

（1）理想设计
（1）**Ideal design**

①屋顶花园可以为高密度环境地面无绿地的机构提供绿洲、增加绿量；

①The construction of roof gardens can provide an oasis for institutions in high-density environments without green areas and increase the amount of greenery；

②充分利用屋顶空间；

②Make full use of roof；

③为机构内部的人员提供安全和私密的空间；

③The roof garden can provide safe and private space for the personnel within the institution；

④只要设计足够安全，就可以为那些长期的住院患者(失智症患者等)提供合适的户外空间；

④As long as the design is safe enough，suitable outdoor spaces can be provided for the long-term hospitalized patients (patients with dementia，for instance)；

⑤可眺望周边的风景，以帮助患者分散注意力、减少压力。

⑤The design should enable the patients to look around and enjoy the view，which will help distract themselves and reduce stress.

(2)非理想设计

(2)Non-ideal design

①由于屋面结构限制，不能使用大树以及具有大荷载的水景和种植池等；

①Due to the limitations of the roof structure，big trees，overloaded waterscapes and planting beds cannot be used；

②在医院里，如果屋顶下面存放关键设备，可能会限制开发屋顶花园，以免漏水；

②In the hospital，if key equipment is stored under the roof，a roof garden may not allowed to develop in order to avoid a water leakage；

③屋顶高度暴露，可能比地面空间或者封闭的庭院更容易遭受风的侵袭；

③The roof is highly exposed and may be more susceptible to wind damage than the ground space or enclosed courtyard；

④邻近建筑物的高度和朝向不同，形成不同的阴影区，使得屋顶花园的温度存在差别，炎热或者寒冷，令人不适；

④Different heights and orientations of adjacent buildings may form different shadow areas. As a result，the temperature of the roof garden can be somewhat different，hot or cold，which may make people feel uncomfortable；

⑤屋顶上设置的暖气/空调装置容易造成噪声干扰；

⑤The heating/air conditioning equipment installed on the roof is prone to generating noise；

⑥需要设置清晰的标识，以引导来访者和患者使用屋顶花园。

⑥Clear signs need to be set up，otherwise visitors and patients would not know the existence of the roof garden.

6.2.3.8　屋顶露台
6.2.3.8　Roof terrace

屋顶露台与屋顶花园不同，屋顶花园位于建筑顶面，而屋顶露台位于建筑的一侧，形成一个狭长的"阳台"。

A roof terrace is different from roof garden. The latter is located on the top of a building, while the former is located on one side of a building in the form of a long and narrow "balcony".

(1)理想设计

(1) Ideal design

①充分利用空间；

①The roof terrace can make full use of space;

②扩展视野，浏览医院外的风光；

②It can broaden people's horizon and enable them to appreciate the scenery outside the hospital;

③在露台内、建筑边缘种植绿色植物，形成迷人的绿色景观，为建筑内部和使用露台者创造一个隐私屏障；

③Green plants can be planted in the terrace and on the edge of the building to form a charming green landscape which can create a natural screen to separate the inside from the outside and protect the privacy of users;

④狭长的空间适合开展运动；

④The narrow and long space is suitable for sports;

⑤当露台位于主门厅或等候区，无须设置标识牌以引人注目，等候区的人很容易发现绿色的露台和远处的景色，可达性好。

⑤When the terrace is located in the main hall or waiting area, there is no need to set up signs to attract attention, because people in the waiting area can easily find the green terrace and see the distant scenery.

(2)非理想设计

(2) Non-ideal design

①露台下方空间如果有较好用处，则不宜修建大露台；

①If the space under the terrace is of great use, do not build a large terrace above it; otherwise it will affect the use of the space below;

②由于结构的限制，露台可能无法种植大树，导致缺少必要的树荫；

②Due to structural limitations, large trees cannot be planted on the terrace, which leads to a lack of necessary shade;

③露台的某些位置可能过冷、过热或者多风；

③The terraces in some locations may be too cold, too hot or windy;

④除非精心设计，有些露台可能会侵犯临近房间的隐私。

④Unless carefully designed, some terraces may be forbidden to use for their invading the privacy of neighboring rooms.

6.2.3.9　中庭花园
6.2.3.9　Atrium garden

中庭花园包括建筑物有玻璃屋顶的天井中的花园，或者阳光房里的花园。在户外气候特别炎热或者寒冷的区域，一年中大部分时间不适合户外休息或者散步，但配备暖气或者空调设备的中庭花园很适合人们休息。

Atrium gardens include those in patios with glass roofs or in sun rooms. In areas where the outdoor climate is particularly hot or cold, it is not suitable for rest or walking outside most of the year, but the atrium garden equipped with heating or air conditioning equipment is very suitable for people to have a rest.

（1）理想设计
（1）Ideal design

①模拟"户外"绿色体验，即使在恶劣天气下也可以使用；

①Simulating the "outdoor" green environment, the atrium garden can be used even in severe weather;

②可达性高，提供了一个安全的、与建筑融合一体的空间；

②The high accessibility of the atrium garden provides a safe space integrated with the building;

③根据空间的大小确定植物体量，即使较高楼层也能欣赏花园美景；

③The density and volume of plants can be determined based on the size of the space, and people on higher floors can easily enjoy the beautiful scenery of the garden;

④为建筑物内部提供充足采光。

④It can provide sufficient light for the interior of the building.

（2）非理想设计
（2）Non-ideal design

①维护植物的室内环境需要更高的成本；

①Maintaining the indoor habitat for plants will cost more;

②为避免养护困难可能使用人造植物，从而降低植物减压的效果；

②To avoid maintenance difficulties, artificial plants may be used, which can reduce the stress-relieving effects of plants;

③由于生长条件不允许，可能使用的人造植物会导致缓解压力的效果不明显。

③Due to different growth conditions, the use of artificial plants may be required, which would reduce plants' effect of relieving pressure.

6.2.3.10　纯观赏花园
6.2.3.10　**Viewing garden**

在空间和预算有限的情况下，一些医院可能会采用一种小尺度的、人不能进入的花园，这种花园只能从建筑内部观看。

In the case of limited space and budget, some institutions may introduce a small inaccessible garden that can only be viewed from the inside of the building.

（1）理想设计
（1）Ideal design

①增加建筑内部采光；

①The viewing garden adds daylight to the interior of a building；

②从相邻的座位或交通空间引向绿色空间视野；

②The neighboring seats or traffic space lead to the view of green space；

③人们可以不被风吹雨淋，就能在室内座位上观赏到室外的绿色空间；

③People can enjoy the outdoor green space on indoor seats without being exposed to wind and rain；

④与能够进人的空间相比，这种纯观赏性的花园维修费用可能更低。

④The maintenance cost of a viewing garden may be lower compared with that of the accessible space.

（2）非理想设计
（2）Non-ideal design

①不能近距离观察、触摸或嗅闻自然要素，多感官感受品质降低；

①As people cannot observe, touch or smell the natural elements at close range, the multi-sensory quality will be compromised；

②无法从室内聆听鸟鸣和水声；

②Even if there are waterscapes or birds in the viewing garden, people indoors cannot hear them；

③无法满足游园的需求。

③It cannot satisfy people's need to stroll, take a walk, or sit idly.

总之，根据医疗建筑的功能和形式不同，可能会出现上述多种户外空间类型，景观设计师需要结合具体项目空间类型，进行深入的思考和分析，综合考虑各类使用群体的具体需求，以及不同地域下的气候条件，进行科学合理的设计。

In short, there may be many types of above-mentioned outdoor spaces according to different functions and forms of medical buildings. Landscape architects need to think deeply and make analysis according to space types of specific projects and comprehensively consider the specific needs of various user groups as well as climatic conditions in different regions to carry out scientific and reasonable design.

思考题

1. 医疗设施康复花园的使用人群有何特点？
2. 医院康复花园的设计要点是什么？

Questions

1. What are the characteristics of the users of healing gardens in medical facilities?
2. What are the key points of designing a healing garden in a hospital?

第 *7* 章

园艺康健花园设计及案例

Chapter 7　Designs and Cases of Horticultural Therapy Gardens

园艺康健花园作为一种典型的康复花园，在欧美及日本较为常见。其名称主要源自用户要在花园中开展园艺活动的初衷，所以园艺康健花园在设计上更偏向于满足园艺活动实施的场地功能性设计。本章以两个日本著名的园艺康健花园为案例展开详细介绍。

As a classic type of healing gardens, the horticultural therapy garden is common in Europe, America and Japan. Its name is mainly derived from the original intention of the users to carry out gardening activities in the garden, so this function is prioritized in the design of a horticultural therapy garden. This chapter introduces in detail the two most famous horticultural therapy gardens in Japan.

7.1　式场医院园艺康健花园

7.1　The horticultural therapy garden of Shikiba Hospital

7.1.1　历史沿革

7.1.1　History

式场医院前身是著名精神病学家式场隆三郎于 1936 年(昭和十一年)开设的精神病医院国府台医院。式场医院位于日本千叶县市川市的国府台之丘，这里距矢切车站五分钟步行距离，是一片树木葱郁风景秀丽的城市绿地，为园艺康健花园的建设提供了良好的环境基础。创始人式场隆三郎非常重视医院环境美化，认为医院不应当让人感到阴暗，院内应当明亮轻松。式场隆三郎同时也是一位具有良好艺术修养的收藏家，参与了日本早年民艺运动，并对梵高的绘画艺术有深入研究。式场隆三郎打破当时医院设计常规，开创性地将医疗建筑与室外花园结合在一起。他认为自然、文学和艺术可以帮助患者恢复心理健康、增强治疗效果。因此将在花园内进行园艺活动作为式场医院精神病治疗的一个重要环节，是日本园艺康健、康复花园的鼻祖之一。为此，式场医院花园的设计方案同时考量了康复景观的视觉艺术性，和在医院开展园艺康健所需的实用功能性。根据数十年累积的园艺康健治疗经验和数据，如今式场医院在保持与开设基本相同空间结构的同时，进行了现代化改造，以适应时代进步（图7-1）。

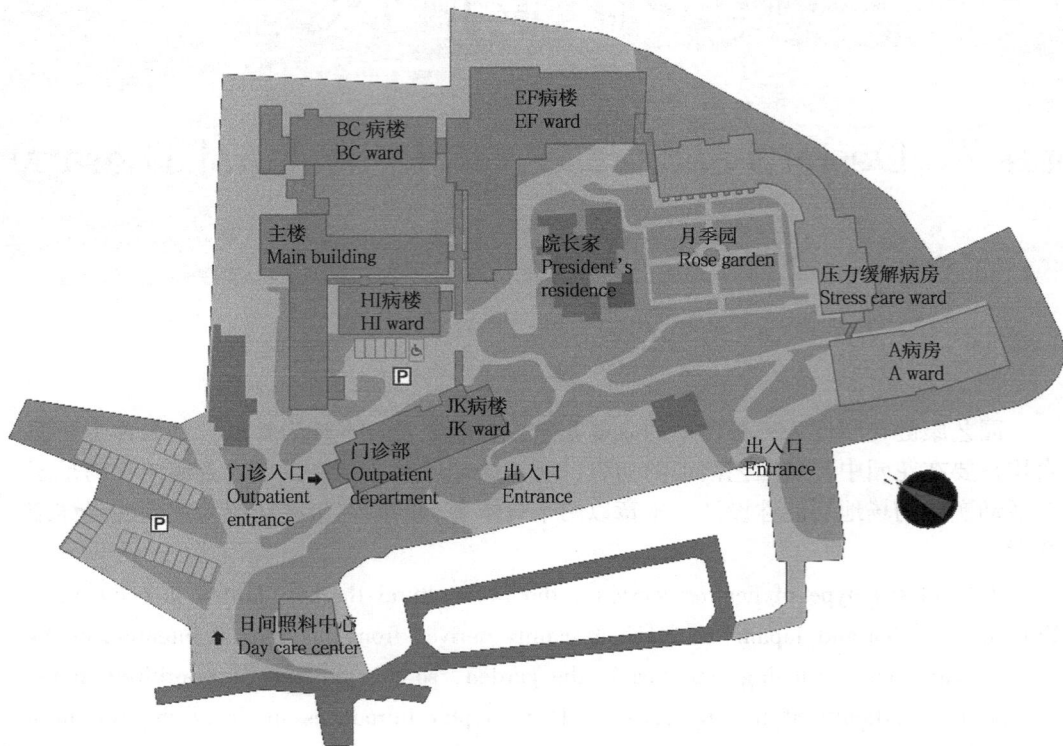

图7-1　式场医院平面配置图（孙旻恺供图）

Fig. 7-1　Floor Plan of Shikiba Hospital（by Sun Minkai）

The predecessor of Shikiba Hospital was Kohnodai Hospital, a psychiatric hospital established by the famous psychiatrist Ryūzaburō Shikiba in 1936 (11th year of the Showa era). Shikiba Hospital is located at a small hill in Konodai, a 5-minute walk from the Yagiri Station in Ichikawa City, Chiba Prefecture, Japan. The small hill in Konodai is an urban green area with lush trees and beautiful scenery, providing a good environmental foundation for the construction of the horticultural therapy garden. The founder Ryūzaburō Shikiba attached great importance to the beautification of the hospital, and believed that the hospital should not be a dark environment but a bright and relaxing one. Shikiba is also a collector with good artistic accomplishment. He participated in the early Japanese folk art movement and made in-depth research on Van Gogh's painting art. Shikiba discarded the conventional hospital design at the time and pioneered the integration of medical buildings and outdoor gardens. He believes that nature, literature, and art can help patients recover their mental health and enhance the effect of treatment. Therefore, gardening activities in the garden are set as an important part of psychiatric treatment in Shikiba Hospital, and he is regarded as one of the originators of horticultural therapy and healing gardens in Japan. For this reason, the visual artistry of the healing landscapes and the practical functionality required for the horticultural therapy in the hospital are considered in the design of the garden in Shikiba Hospital. Based on the experience and data of horticultural therapy accumulated over decades, Shikiba Hospital has been modernized to adapt to the advancement of the times while maintaining the basic spatial structure when it was founded (Fig. 7-1).

7.1.2　花园布局
7.1.2　The garden layout

　　式场医院总占地 21 280m²，其中医院总面积约一半即 9900m² 为室外花园。从平面图（图 7-1）可以看到，式场医院拥有一个占地极大的月季园，其主要治疗建筑均以该花园为中心环绕布局。离月季园稍远处的建筑则专门配有中庭，以确保从每一栋建筑内向外观望时都可以见到绿色景观。月季园中约有 600 种 2000 株月季。式场隆三郎曾撰文称："对医院来说，环境非常重要，尤其是精神医院，可说事关重大……比起医院室内环境的改善，我希望医院的院长或是经营者更加重视医院内部庭院的改善。"从式场医院的平面构成来看，式场医生彻底执行了自己的理念，特别是针对由压力过大而引起心理健康问题的患者，他们将病房建在了花园的最佳观景地点。该建筑位于月季园东侧，名为"月季别墅"。月季别墅的弧线外形设计保证从每一个房间都能将月季园尽收眼底，体现了通过景观来辅助治疗患者心理问题的方针（图 7-2）。

　　Shikiba Hospital covers a total area of 21 280m², of which the outdoor garden occupies 9900m², almost half of the total area of the hospital. The floor plan (Fig. 7-1) shows that Shikiba Hospital has a huge rose garden, and its main buildings for treatment are located around the garden. In the buildings a little farther away from the rose garden, atriums are built to ensure that the green landscape can be seen from inside of each building. There are about 2000 roses of 600 species in the rose garden. Ryūzaburō Shikiba once wrote, "For hospitals, the environment is very important, especially for mental hospitals. It is of great significance … Compared with the improvement of the hospital's indoor environment, I hope that its director or operator will pay more attention to the improvement of its inner courtyard. " The layout of Shikiba Hospital has shown that doctors of this hospital have thoroughly implemented their own philosophy. The stress relief wards for patients with mental health problems caused by excessive stress are in a building at the best spot for viewing in the garden. The building is located on the east side of the rose garden and is named "Rose Villa". The design of its curved shape ensures that the rose garden can be seen from every room, which embodies the policy of assisting the healing of psychological problems with landscape (Fig. 7-2).

图 7-2　式场医院月季园及"月季别墅"（孙旻恺　摄）
Fig. 7-2　Rose garden and "Rose Villa" in Shikiba Hospital（by Sun Minkai）

　　为了在维持开放式花园设计理念的同时，满足精神病院的特殊使用需求，式场医院非常注重景观设计和配置。首先，在患者入住的"月季别墅"花园对面，是院长家的住宅(图 7-3)，以便于院长随时了解患者情况，而病房与院长住宅之间有花园缓冲，亦不会使住院患者心生被监视的疑虑。另外，早期式场医院为了消除精神病院闭塞阴郁的印象，去除了所有的外围栅栏(现在医院外周有一部分围栏，隐藏在月季花丛和其他植物之后，从医院内部无法看见)。在没有围栏的条件下，如何防止患者在治疗期间离开医院，成了需要解决的问题。式场医院则将医院连接前后门的通路路线设计成曲折形，隐藏在月季园花丛之后，颇为曲幽，从医院内侧无法直观判断出入口在哪里。这样既可防止患者从医院走失，又降低了空间的闭塞感和压迫感(图 7-4)。

　　In order to meet the special needs of the mental hospital while keeping to the design philosophy of an open garden, Shikiba Hospital examined and weighed the landscape design and plant arrangement thoroughly. First, opposite to the "Rose Villa" is the director's house (Fig. 7-3), so that the director can observe the patients' status at any time. The garden also serves as a buffer zone between the building and the director's house, easing hospitalized patients' worry about being monitored. In addition, in the early phase, in order to eliminate the isolated and gloomy atmosphere of mental hospitals, all Shikiba Hospital's outer fences were removed, and now there are some fences in the periphery of the hospital, hidden in the rose bushes and other plants, so they cannot be seen from inside the hospital. As a result, without fences how to prevent patients from leaving the hospital during treatment has become a problem that needs to be solved. The approach of the Shikiba Hospital is to design the road from the front door to the back door into a zigzag shape, which, rather twisting, is hidden behind the flowers of the rose garden, so it is impossible to directly figure out where the entrance and exit are inside the hospital. This approach not only prevents the patients from getting out of the hospital, but also reduces the sense of seclusion and pressure in the space (Fig. 7-4).

图 7-3　式场医院月季园及院长宅（孙旻恺　摄）

Fig. 7-3　Rose garden and director's house in Shikiba Hospital（by Sun Minkai）

图 7-4　式场医院出入口周边（孙旻恺　摄）

Fig. 7-4　Entrance of Shikiba Hospital（by Sun Minkai）

7.1.3　园艺活动和植物配置
7.1.3　Gardening activities and plants arrangement

　　月季园不仅通过视觉、嗅觉等感官刺激发挥康复作用，也是实施园艺康健辅助精神病治疗的主要活动场所。式场隆三郎称"绿化必要性并不只体现在一个方面，除了让患者感到赏心悦目之外，让患者参与园艺活动也是目的之一。庭院、花圃作为园艺康健的绿化基础，越发显得重要"。式场医院的患者都会获得参与花园管理工作的机会。式场医院职员 Nakanisishun 详细描述了当时进行园艺活动的情形，"想要月季花盛开，每日都需要勤加管理。约有 10 名患者每日与指导者（责任人及其徒弟）早早地外出打理，于是其他症状稍重的患者也希望能够外出，只盼何时能轮到自己外出打理月季。同时，让患者参加打理月季也是测试患者治愈程度的一种手段"。

　　The Rose garden exerts its healing effects through sensory stimuli such as sight and smell, and it is also the main site for the implementation of horticultural therapy-assisted mental treatments. Ryūzaburō Shikiba said, "The necessity of green space is reflected not only in gladdening patients' heart and pleasing their eyes. Encouraging patients to participate in gardening activities is also one of the purposes. Courtyards and parterres, as the greening basis of horticultural therapy, are becoming more and more important." Patients in Shikiba Hospital would be given the opportunity to participate in garden management. A staff member surnamed Nakanishun in Shikiba Hospital described in detail the gardening activities at that time, "If you want the roses to bloom, you need diligent management every day. About 10 patients go out early to take care of those roses daily with instructors (the attending therapist and his apprentices). Then other patients with severe symptoms also ask to go out, eager for their turn to take care of roses. So participating in roses management is also a way to test the degree of patients' recovery".

月季盛开时节，医院对公众开放，每日游人接踵而来，尤其在春季和秋季，医院附近幼儿园、小学的孩童会来医院花园赏花郊游，在草坪上野餐玩耍。此外，医院花园也成为附近居民散步的去处。像这样附近居民可以随意进出医院花园的开放式设计，使得人们在赏花的同时也不经意间观察到精神病患者，在确保安全的前提下会有简单交流，这对帮助精神病患者融入正常社会，以及消除一般群众对精神病患者的偏见，具有很好的作用。

When the roses are in full bloom, the hospital is open to the public, and visitors come every day, especially in spring and autumn. Children from kindergartens and elementary schools near the hospital usually go to the hospital garden to enjoy flowers, and have picnics on the lawn. In addition, the hospital garden has become a place where local residents can stroll. Such an open design allows visitors enter and exit the hospital garden at will, and they may inadvertently observe mental patients while admiring the flowers, and even have simple exchanges with them when it is safe to do so. It has played an excellent role in helping mental patients integrate into society and eliminating the general public's prejudice against them.

式场医院实施的园艺康健方案与月季园设计相辅相成。首先，月季园采用几何式设计，其特征在于花园内月季及其他植物规整种植，构成几何图案，展现规整之美。为达到最佳观赏效果，植物修剪整理等园艺作业必不可少。其次，月季为开花灌木，为使月季达到最佳开放效果，必须勤加施肥、经常修剪。月季是一种小型灌木，郁闭度较低，由于施肥勤，较易滋生杂草，同时病虫害较多。综合来说，若需维持月季园的美观需要较大强度的养护管理，这为园艺活动的经常性开展提供了条件。与此同时，月季枝条上长有尖刺，若作业不小心，则有被刺痛的危险，因此迫使参加园艺活动的患者集中精力，冷静地进行细致作业，锻炼患者保持平和心态的情绪控制能力。医生与患者共同种植的月季，除供花园观赏之外，也做切花等其他用途。"我院创立以来，已经定下方针，勿令患者从事有盈利性的活动，而我等培育的月季花都会无偿赠予爱好者。我们最为开心的时刻，莫过于患者治愈出院之时，带着自己亲手栽培的月季花与家人一同回家，简而言之，月季之于我等亦等同于一剂良药。"

The horticultural therapy plan in Shikiba Hospital and the rose garden design are supplementary to each other. First, the rose garden adopts a geometric design, which is characterized by the orderly planted of roses and other plants in the garden, showing the beauty of regularity. In order to present the best viewing effect, gardening activities such as plant pruning and trimming are indispensable. Second, roses are flowering shrubs. If you want to achieve the best blooming effect, you must fertilize and prune frequently. Rose is a small shrub with a low canopy density, and frequent fertilization will breed weeds, and at the same time cause more pests and diseases. Overall, if you need to maintain the beauty of a rose garden, you need relatively intensive maintenance and management, which creates the need for regular gardening activities. As there are sharp thorns on the rose branches, if you work carelessly, you are in danger of being pricked. Therefore, patients who participate in gardening activities have to concentrate and work carefully, enhancing the emotional control ability to maintain a calm mind. The roses planted by doctors and patients are used not only for viewing in the garden, but also for other purposes such as cut flowers. "Since the establishment of the hospital, we have set a policy of not allowing patients to engage in activities with monetary gains, and the roses we cultivated will be given free of charge. There is no happier moment than when cured patients go home with their family taking the roses they cultivated themselves. In a word, the roses are a kind of medicine to us."

7.1.4 效果总结
7.1.4 A summary of effects

式场医院可谓以康复景观为中心构筑了整个医院的空间结构，结合景观视觉刺激的康健作用，使用庭院进行园艺康健的实践，开放庭院改善了公众对精神病患者的认知。种种尝试至今仍具先进性，从理念到实践、从宏观到微观，都为后来者做出了很好的榜样。

The spatial structure of Shikiba Hospital was built with the healing landscape as the center; in combination with the healing effect of visual stimulation of the landscape, the hospital uses the courtyard to practice horticultural therapy and opens the courtyard to improve the public's understanding of mental patients. These attempts seem very advanced even today. From concept to practice and from macro to micro, the hospital has set an excellent example for the latecomers.

7.2　浜野医院园艺康健花园
7.2　The horticultural therapy garden of the Hamano Hospital

7.2.1　历史沿革
7.2.1　History

浜野医院是创设于 1950 年的综合性医院。医院位于日本长崎县佐世保市俵町，佐世保市是一座四面环海的港口城市，市区围绕两座山体建成，气候温润、景观秀丽。医院所在地俵町位于市内山脚下一处较为繁华的地区。其主体是两座大楼，是一所结合门诊与住院的综合性医院。受限于医院所处位置，浜野医院占地面积极小，几乎没有室外空间可供使用，对设置园艺康健花园来说，其基础环境条件并不理想。随着日本社会老龄化程度不断加重，医院主要服务人群为俵町周边居民，医院来访者大部分是高龄患者。因此该医院主要应对患有慢性病的老年人。随着园艺康健在日本的普及开展，其对维系老年人身心健康、提高生活质量的有效性得到了广泛肯定。因此浜野医院自开设以来，就在医院内开展多种形式的园艺康健。医院顶层天台分别改造为屋顶菜园和屋顶庭院，用于不同园艺康健项目（图 7-5、图 7-6）。

图 7-5　浜野医院(孙旻恺　摄)

Fig. 7-5　Hamano Hospital（by Sun Minkai）

	7	天空花园 Sky garden
屋顶菜园 Rooftop vegetable garden	6	会议室 Meeting room
若叶组团 Group home WAKABA	5	若叶组团 Group home WAKABA
疗养病房 Convalescent ward		
社区病房 Community ward		
康复病房，手术室 Rehabilitation ward，Operating room	2	日间照料，浴室 Day care，Bath room
办公室 Office	1	门诊部 Outpatient department

图 7-6　浜野医院楼层示意图(孙旻恺供图)

Fig. 7-6　Floor plan of Hamano Hospital（by Sun Minkai）

Hamano Hospital is a general hospital established in Tawaramachi, Sasebo City, Nagasaki Prefecture, Japan in 1950. Sasebo City is a port city surrounded by the sea and its urban area, with a mild climate and beautiful landscape, is built around two mountains. The Tawaramachi where the hospital is located is a downtown area at the foot of a mountain in the city. With two buildings as its main body, the hospital is a general hospital that provides both outpatient and inpatient services. Limited by the location of the hospital, it covers a very small area and almost no outdoor space can be used. The basic environmental conditions are not ideal for setting up a horticultural therapy garden. The main population served by the hospital are residents in the surrounding areas of Tawaramachi, and due to Japan's aging society, most of the hospital visitors are elderly patients. Therefore, the hospital mainly deals with elderly people with chronic diseases. With the implementation and popularization of horticultural therapy in Japan, its effectiveness in maintaining physical and mental health of the elderly and improving their quality of life has been widely recognized, so various forms of horticultural therapy have been carried out in the hospital since its establishment. The rooftop of the hospital has been transformed into a vegetable garden and a courtyard for different horticultural therapy programs (Fig. 7-5, Fig. 7-6).

7.2.2　屋顶菜园布局
7.2.2　The layout of the rooftop vegetable garden

　　位于第六层屋顶的屋顶菜园,其主要园艺种植活动参与人群,是具有一定身体活动能力和认知能力的患者,为防止住院患者在楼顶区域出现意外,楼顶周边均设有钢架并嵌入玻璃,在保证安全的同时可以从屋顶眺望周边景观。这一区域同时还兼有晾晒衣物被褥等用途,因此在其出入口附近设有一处较为宽阔的硬质铺装晒台。晒台不仅用于晾晒衣物,在使用可移动种植设备进行园艺活动时也可以作为活动场地(图7-7)。在靠近出入口一侧的钢制顶棚上装有玻璃,以保证阴雨天也能顺利开展活动。入口处的右侧设有花坛,花坛中种植四季花卉,花卉的品种选择与管理则皆由参加园艺康健项目的患者实施。花坛为方便老年人进行活动,围边设置较一般花坛高,以同时满足弯腰进行园艺活动的老年人和使用座椅、轮椅进行园艺活动的老年人的需要。同时,为方便园艺活动时护理人员从旁协助,花坛的形状设置为锯齿形(图7-8)。折线形状的花坛使得相邻的两位参与者侧面相对,相比于参与者排成一列进行活动更方便互动,可促进参与者之间的交流。花坛的对侧设有椅子和桌子,方便园艺活动参与者或其他来访人员坐下观赏。同时,在园艺活动参与者较多的情况下可为部分等候的参与者提供休息场地。

图 7-7　浜野医院屋顶菜园(孙旻恺　摄)
Fig. 7-7　Rooftop vegetable garden in Hamano Hospital (by Sun Minkai)

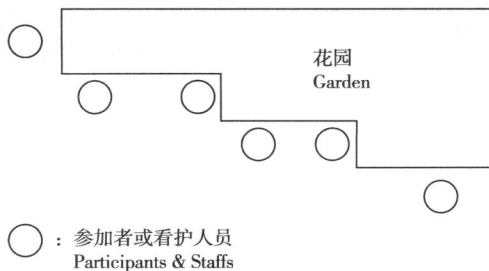

花园
Garden

◯ ：参加者或看护人员
Participants & Staffs

图 7-8　花坛平面图(孙旻恺供图)
Fig. 7-8　Plan of flower bed
(by Sun Minkai)

First, the main participants of the horticultural activities in the rooftop vegetable garden on the sixth floor are the patients with a certain degree of body movement and cognitive abilities. In order to prevent inpatients from accidents on the roof, steel frames, is embedded with glass, are set up around the roof to ensure safety without obstructing the view of the surrounding landscape. This area is also used for drying clothes and bedding, so there is a wide flat hardscape platform near the entrance. The flat platform can be used not only to dry clothes, but also as a site for gardening activities when mobile planting equipment are used (Fig. 7-7). Glass is installed on the side close to the entrance of the steel frame to make activities possible even in cloudy or rainy days. A flowerbed is set up on the right-hand side of the entrance with flowers planted at all seasons seasons. The varieties selection and management of flowers and plants are done by the patients participating in horticultural therapy programs. In order to facilitate the activities of the elderly, higher flowerbeds are also set up to meet the needs of both the elderly who bend over and those who use chairs and wheelchairs for gardening activities. At the same time, in order to facilitate the nursing staff in assisting the participants in gardening activities, the shape of those flowerbeds is set to be zigzag (Fig. 7-8). The zigzagging flower beds make two adjacent participants face each other sideways, so compared with lining up in a row, it is easier to promote communication among participants. Chairs and tables are put on each side of a flower bed, so that gardening participants or other visitors can sit and watch. A rest place is provided for waiting participants when there are too many participants in gardening activities.

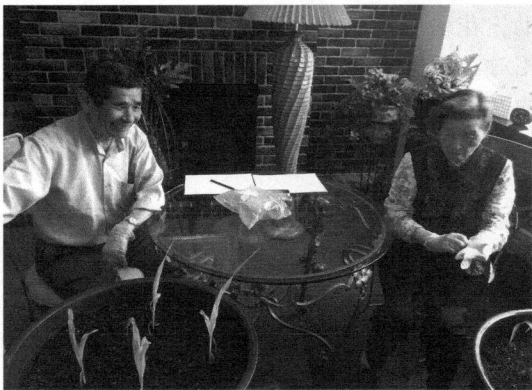

图7-9 大型容器内的玉米幼苗(孙旻恺 摄)

Fig. 7-9 Maize planted in large containers

(by Sun Minkai)

由于屋顶花园无法提供足够的土壤环境，为了种植某些与其他疗法结合效果较好(尤其是针对失智老人的康复训练，如算数疗法等)的蔬果品种如玉米、草莓等，花园配套设置了大型种植容器，将这些作物种植在内，并固定在推车上以方便移动(图 7-9)。

As the roof garden cannot provide enough soil to plant some fruit and vegetable varieties such as corn, and strawberry whose combination with other therapies (especially the rehabilitation training for the elderly with dementia, such as arithmetic therapy) delivers better effects, large containers for planting these crops are laid out in the garden and fixed on carts for easy mobility (Fig. 7-9).

晒台一旁设有菜园区块，受限于原建筑结构，这一区块地面低于晒台，因此在两者之间设置了无障碍通道。通道的宽度为 1m，使用轮椅或步行辅助器的老人可以顺利通过，也方便搬运园艺器材。在无障碍通道的两侧设有栏杆，帮助参与者行动。菜园区块内主要由三处种植区、一处休息区和一处工具材料存放区构成。三块种植区分别种植小型果树、蔬菜瓜果和香草。休息区位于最大的种植区块：蔬菜瓜果类的斜对方，方便在各个区域进行园艺活动的参与者休息（图7-10、图7-11）。

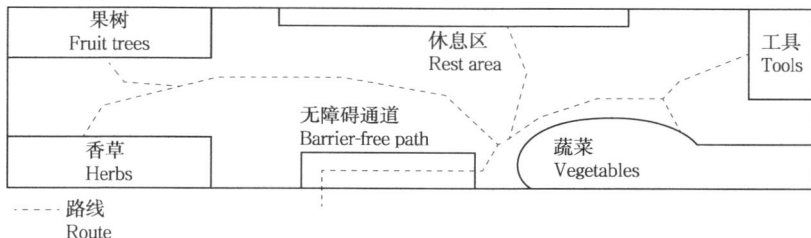

图 7-10　屋顶菜园平面（孙旻恺供图）
Fig. 7-10　Plan of rooftop vegetable garden（by Sun Minkai）

There is a vegetable garden beside the flat platform. Limited by the original building structure, the area is lower than the flat platform, and a barrier-free passage is set between the vegetable garden area and the flat platform. The width of the passage is 1m, enabling the elderly using wheelchairs or walking aids to pass smoothly and making it easier to carry gardening equipment. There are railings on both sides of the barrier-free passage to help participants move. The vegetable garden area is mainly composed of three planting areas, a rest area and a tool and material storage area.

图 7-11　老人在参加园艺活动（孙旻恺　摄）
Fig. 7-11　A senior citizen is participating in horticulture therapy（by Sun Minkai）

Small fruit trees, vegetables, melons, and herbs are planted in the three planting areas. The rest area is diagonally opposite to the vegetables and fruits area, in the largest planting area, making it convenient for the participants in each area to rest（Fig. 7-10, Fig. 7-11）.

7.2.3　屋顶菜园园艺活动和植物配置

7.2.3　The gardening activities and plant configuration of the rooftop vegetable garden

由于屋顶菜园区域面积较大，在进行园艺活动时需要进入种植坛内进行作业，因此与进深较浅的花坛相比，此处高度较小，方便参与者进出。为配合园艺康健疗程，每个区域的植物都按季节种植及采收。从锻炼身体技能角度出发，蔬菜瓜果区主要种植三大类作物配合三类园艺活动。

As the rooftop vegetable garden covers a large area, it is necessary to enter the planting beds to work during the gardening activities. Therefore, compared with narrower flower beds the planting beds here are relatively lower to allow easy access. In coordination with the horticultural therapy treatment, the plants in each area are selected according to seasons and planting and harvesting methods. From the perspective of improving physical prowess, three types of crops are mainly planted in the vegetable and fruit area for three kinds of gardening activities.

①需要精细手部操作进行采收的蔬菜，如番茄、茄子、秋葵、菜椒等。

①Vegetables to be picked with delicate manual operation such as tomatoes, eggplants, okra and bell peppers.

这些蔬菜株高大多在50~100cm，果实结果时期较长，果实较小，数量较多，适合老年人进行采摘作业等，以此可锻炼视力和手部运动能力、协调能力。

The height of these vegetables varies from 50 to 100cm. With a long fruiting period, small fruit and a large quantity, those vegetables are suitable for the elderly to pick to improve their eyesight and hand movement and coordination abilities.

②需要使用较多体力进行收获的蔬菜，如洋葱，生姜等。

②Vegetables to be harvested with more physical strength such as onions and ginger.

这些蔬菜的食用部分为植物的根茎部分，收获时需要用较大的力气将其从土壤中拔出，可以借此锻炼参加者的臂力、腰腿部力量。

The edible parts of those vegetables are their roots and stems that need to be pulled out of the soil with greater strength when being harvested, so such an activity can improve the arm, waist and leg strength of the participants.

③藤蔓类蔬菜，如黄瓜、苦瓜等。

③Vine vegetables, such as cucumber and bitter gourd and so on.

此类蔬菜生长在网架之上，采摘瓜果时需要举手抬头，因此用于锻炼参与者的上肢及肩部运动能力。此外，此类蔬菜种植在蔬菜瓜果区域靠墙一侧，可通过挂网向上爬满廊架，在夏季进行园艺活动时遮挡烈日、提供阴凉。

This kind of vegetables grows on a grid, and patients need to raise their hands and head when picking the vegetables and fruits, so such an activity is used to improve the movement ability of the upper limbs and shoulders of participants. In addition, this kind of vegetables is planted on the side, of the vegetable and fruit area, close to the wall, and they can climb up through hanging nets to cover all of the grid, blocking the hot sun and provide shade during gardening activities in summer.

果树区主要种植柑橘类果树，如柠檬、香柑等。这一类植物的果实个体较大、色彩艳丽，适合采摘，并且与前述月季等一样，柑橘类植物的枝干长有尖刺，可以在一定程度上促使参与者集中精力在手部动作的精细控制上，达到康复治疗的效果。另外，柑橘类植物开花时花香浓郁，可对参与者的嗅觉造成良性刺激。从植物种植角度看，柑橘类植物抗逆性强，可以充分适应楼顶土壤贫瘠、风速大的恶劣环境，适合在楼顶种植。

Plants in the fruit tree area are mainly citrus trees, such as lemons and citruses. The fruits of this type of plant are relatively large, colorful and suitable for picking. Like the aforementioned roses and Chinese roses, there are thorns on the branches of citrus plants, thus making participants to concentrate on the fine control of hand movements to a certain extent to achieve the effect of rehabilitation. In addition, citrus plants have a strong floral fragrance when they bloom, which causes a benign stimulus to the participants' sense of smell. From the perspective of planting, citrus plants are highly stress-resistant and can fully adapt to the harsh environment of the barren soil and high wind speed on the roof, suitable for rooftop gardens.

香草区主要种植薄荷、迷迭香、罗勒等可食用香草类。香草的香味对参与者的嗅觉刺激有助于改善参与者的心理状态，调节情绪。收获后的香草可用于烹调或作为工艺加工原材料。并且香草类植物大多抗逆性强，适宜用于屋顶绿化。

Edible herbs such as mint, rosemary, and basil are mainly grown in the herb area. The scent of herbs stimulates the participants' sense of smell and helps to improve the mental state of the participants and regulate their emotions. The harvested herbs can be used for cooking or as raw materials for arts and crafts processing. And most of herbs are highly stress-resistant and suitable for roof greening.

7.2.4　屋顶庭院布局
7.2.4　The layout of rooftop courtyard

位于第七层屋顶的天空花园是一处屋顶庭院。庭院已被证明具有促进观赏者内心平和的作用，特别是对于丧失身体活动能力的老人和失智老人，观赏者只要身处庭院空间之中，即可起到极好的情绪调节作用。进一步研究表明，只有在无阻碍（玻璃）等的情况下直接观赏实体庭院（非照片或影像）才能达到最好的观赏效果。这些研究成果，为无法参加需要身体活动能力才能进行园艺活动的老年人，提供了参与园艺康健的新途径，也为在医养机构设置观赏性绿化景观提供了理论基础。

The sky garden on the seventh floor is a rooftop courtyard. Courtyards hare been proved to have the effect of promoting the inner peace of the viewers, especially for the elderly who have lost physical performance and suffered dementia. Viewing the courtyard or just being in it can play an excellent role in emotional regulation. Further research has shown that the best viewing effect can be achieved only by directly viewing the physical courtyard (not photos or images) without obstruction (glass). Those research results provide a new way to receive horticultural therapy for the elderly who cannot participate in gardening activities that require physical strength, and lay down a theoretical basis for setting up ornamental green landscapes in medical care institutions.

浜野医院七层花园主要为身体活动能力较弱、认知能力有障碍的患者提供景观观赏场地。花园与最上层医护人员多功能食堂相连，从室内对庭院进行观赏是主要的利用方式，这也是庭院的传统观赏模式之一（图7-12）。因此该花园在设计时未在庭院内部设置无障碍通道，以免影响庭院的观赏效果。考虑到此处庭院也位于屋顶，为防止出现意外情况，屋顶四周设置了高 4m 的墙壁。相对于庭院整体面积而言，为防止墙体相对过高产生空间压迫感，因此在墙体上设置窗口以采光。

The garden on the seventh floor of Hamano Hospital provides a place mainly for patients with weak physical performance and cognitive impairments to view the landscape. The garden is connected to the multifunctional canteen for medical staff on the topfloor, and viewing the courtyard from the interior is the main way to use the courtyard. This is also one of the traditional ways to view the courtyard (Fig. 7-12). Therefore, there were no barrier-free passages inside the courtyard in the design of the garden, to avoid affecting the viewing effect of the courtyard. Considering that the courtyard is also located on the roof, in order to prevent accidents, a four-meter wall is built around the roof. Compared with the overall area of the courtyard, the wall is relatively high, and to prevent a sense of spatial pressure, windows are provided on the wall for to allow more light.

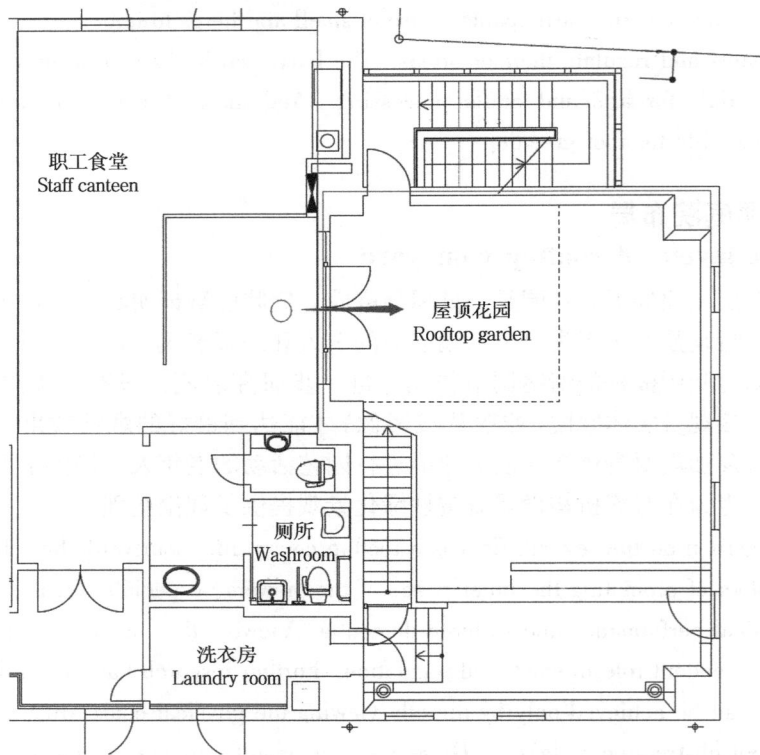

图 7-12　屋顶庭园平面图(孙旻恺供图)

Fig. 7-12　Plan of rooftop courtyard (by Sun Minkai)

7.2.5　屋顶庭院植物配置
7.2.5　The plant preparation on the rooftop courtyard

整个庭院在适应建造环境的基础上，严格按照日式露地茶亭风格进行设计（图 7-13）。庭院最远端设有玻璃窗，可以眺望远处山景，即借远处山峦作为该庭院的远景。此举可增加空间延展性，加强与墙壁相邻一侧则种植竹子，利用竹子弱化墙壁的生硬触感，减轻空间压迫感。而竹子树形修长，适宜种植在狭小空间内。墙脚下则种植各类蕨类植物和小灌木，与竹子、山峦远景共同抽象勾画竹林山野的氛围，这正是日本传统乡村"里山"（里山是指传统日本村落前村后山的结构，浜野医院收治患者以老年人居多，其年轻时多数生活在日本高速城市化之前的传统村落）

图 7-13 屋顶庭园（孙旻恺 摄）
Fig. 7-13 Rooftop courtyard（by Sun Minkai）

的传统空间特征，目的在于勾起观赏者的乡愁，促进记忆回溯，以帮助维持自我认知。左侧墙壁由于结构原因无法增设窗户，因此沿墙壁种植竹子，削减墙壁的压迫感。庭院内主要构成要素为垂枝鸡爪槭、春日石灯笼、水盆、矮灯笼以及汀步和碎石铺地。

On the basis of adapting to the construction environment, the entire courtyard was designed in strict accordance with the Japanese open-field tea pavilion style（Fig. 7-13）. The far end of the courtyard is equipped with glass windows, through which people can overlook the distant mountains. That is to take the distant mountains as the background of the courtyard. The design can expand the space, strengthen the visual connection with the surrounding environment, and eliminate the sense of seclusion caused by the small space. Bamboos are planted along the wall to weaken the rigid touch of the wall and reduce the sense of oppression. The slender bamboos are suitable for small space. At the foot of the wall, various ferns and small shrubs are planted to create the atmosphere of the bamboo groves and mountains together with the bamboos and the distant view of the mountains. This is the exact traditional spatial structure of satoyama of Japanese traditional villages（Satoyama refers to the structure of traditional Japanese villages front of the mountains. The patients treated in Hamano Hospital are mostly the elderly. When they were young, most of them lived in the traditional villages before the rapid urbanization in Japan）. The purpose is to evoke the nostalgia of viewers, promote memory recalling, and help maintain self-cognition. Windows cannot be added to the wall on the left due to structural reasons, so bamboos are planted along the wall to reduce the sense of oppression. The main elements of the courtyard are the weeping Japanese maples, Kasuga stone lanterns, water basins, low lanterns, stepping stones on water surface and rubble pavement.

鸡爪槭种植于庭院西北角后方，每到秋季红叶时，不但赏心悦目，而且能增强参与者的季节认知。整个庭院分为两条主轴线，第一条轴线从观赏者所在地点出发，顺汀步延伸到庭院尽头向东北转而消失在竹篱之后，似通往住户人家；另一条轴线与山峦呼应，又好像直通山脚，此种想象上的延伸所表现的也是日本"里山"原生风景。第二条轴线则是春日石灯笼、水盆、矮灯笼三点连线，其中春日石灯笼与水盆形成主景，位于视觉中心位置稍西以避免中心对称，这也是日本乃至亚洲园林的基本设计方式。同时，两条轴线的设计避免了单点消失型空间结构变化小、不适宜长时间坐观的缺点。水盆上设有惊鹿(一种小型装置，随着水在竹管内积存，平衡改变，竹管一侧跌下敲打在景石上发出声响)，惊鹿为静态景观为主的庭院增加了一份动态景观，增强了庭院对观赏者的视觉刺激，同时，惊鹿发出声音可以提供听觉刺激，从多个角度提供感官刺激，增强庭院的可观赏性。

Japanese maples are planted at the left rear of the courtyard. In autumn, admiring the red leaves is not only pleasing to the eyes, but also enhances the participants' awareness of the seasons. The entire courtyard consists of two main axes. The first axis starts from the place where the viewers are, extends along the stepping stones on water surface to the end of the courtyard, and then turns right and disappears behind the bamboo fence, seeming to lead to the residents' houses. The other axis echoes the distant mountains and seems to go straight to the mountain foot. This imaginary extension shows the original scenery of the Japanese satoyama. The second axis is the three-point connection of the Kasuga stone lantern, water basin, and low lantern. The Kasuga stone lantern and the water basin form the main scene, which is located slightly to the left of the visual center to avoid central symmetry. This is the basic design of Japanese and Asian gardens. At the same time, the design of the two axes avoids lack of variation in the space structure due to only one vanishing point being not suitable for viewing for a long time. The water basin is equipped with shishiodoshi (a small device where a bamboo tube falls on one side and hits on a stone to make a sound as the water accumulates in the bamboo tube to cause a balance change). The Presence of shishiodoshi adds a dynamic scene to the courtyard dominated by static landscapes and enhances its visual stimulation to viewers. What's more, shishiodoshi's sound can provide auditory stimulation, and improve the courtyard's ornamental value by providing sensory stimulations from multiple angles.

7.2.6　效果总结
7.2.6　A summary of effects

老年观赏者在观赏庭院时的眼球运动轨迹表明，设计精巧的庭院比未经设计的屋顶开放空间更能激活观赏者的眼球运动，促进观赏者对庭院的视觉探索，观赏者在观赏庭院景观时，视线分布面广，而非集中在中间区域，证明庭院景观对观赏者具有较强的吸引力，而观赏非庭院空间时，观赏者的视线移动较少，集中在画面中间处，庭院空间无法引起观赏者的兴趣。自主神经活动数据表明，观赏庭院时观赏者的交感神经兴奋(图7-14)。

The eye movement track of elderly viewers when they view the courtyard show that a well-designed courtyard can better activate their eye movement and promote their visual exploration of the courtyard than other undersigned open spaces on the roof. When viewers are viewing the courtyard landscape, their eyesight covers a wide area, rather than focus on the central area, which proves that the courtyard landscape has strongly attracted them, but when they are viewing the non-courtyard space, their sight moves less, focusing on the center of the scenery, which proves that the non-courtyard space fails to cause their interest. Autonomic nerve activity data indicate the sympathetic nerves of the viewers are excited when they are viewing the courtyard (Fig. 7-14).

图 7-14　观赏庭院(左)与非庭院空间(右)时的视线轨迹图(60秒)(孙旻恺　摄)

Fig. 7-14　Comparison of the eye movement in designed garden(left)
with that in non-designed greenspace (right) (60 seconds) (by Sun Minkai)

　　虽然具体原理尚待探明，但在实际运用过程中，浜野医院主要利用该庭院预防老年患者黄昏症候群(又称"日落综合征")发作，且取得了较好效果。日落时分，有的老年人可能会突然发生意识障碍，如情绪紊乱、焦虑、亢奋和方向感消失等，持续时间为几个小时或者整个晚上。这种急性神志紊乱状态，因常发生在黄昏的时候，在医学上称为"黄昏症候群"。黄昏症候群对患者的生活质量造成了严重的负面影响，也加重了看护人员的负担。在浜野医院，每当医护人员判断患者有黄昏症候群发作征兆时，会陪伴患者至该花园进行观赏，这样能有效改善患者的情绪，遏制黄昏症候群发作率(图 7-15)。

Although the specific principles have yet to be explored, in the actual application process, Hamano Hospital mainly uses the courtyard to prevent the onset of sundown syndrome in elderly patients. At sunset, some old people may suddenly experience disturbances of consciousness, such as mood disorders, anxiety, excitement and losing sense of direction, which last for a few hours or throughout the night. This kind of acute mental disorder is called "sundown syndrome" in medicine because it often occurs at dusk. Sundown syndrome has a serious negative impact on the quality of life of patients, and it also increases the burden on caregivers. At Hamano Hospital, whenever medical professionals judge that the patient are suffering from sundown syndrome, they will accompany the patient to see the garden, which has effectively improved the patient's mood and reduced the incidence of sundown syndrome (Fig. 7-15).

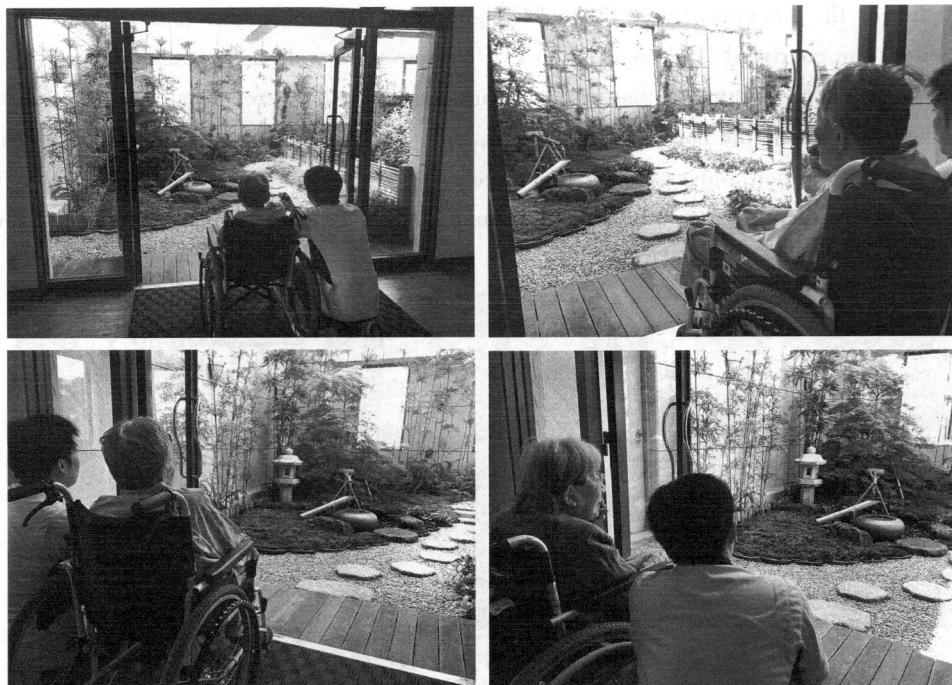

图 7-15 老人在护工陪同下观赏庭院(孙旻恺 摄)

Fig. 7-15 Residents are viewing garden with hospital staff (by Sun Minkai)

思考题

请结合案例归纳园艺康健花园的设计特点。

Question

Please summarize the design characteristics of horticultural therapy garden based on case studies.

第 8 章

阿尔茨海默病（AD）患者康复花园特点及案例

Chapter 8 Characteristics and Cases of Healing Gardens for Patients with Alzheimer's Disease (AD)

8.1 AD 专类园概念及特点

8.1 The concept and characteristics of AD gardens

阿尔茨海默病（以下简称 AD），是最常见的失智症类型。临床上以记忆障碍、失语、失用、失认、视空空间感损害、执行功能障碍，以及人格和行为改变等全面性失智表现为特征。AD 患者一开始主要表现为记忆衰退、空间方位感差、易患幽闭恐惧症并伴有梦游倾向，逐渐发展为认知功能障碍，甚至神志不清、瘫痪在床。研究表明，AD 是造成老年人失去日常生活能力的最常见疾病，同时也是导致老年人死亡的第五位病因，当前还没有失智症彻底治愈的方法，但是可以通过早期开展认知功能和体力锻炼，来延缓 AD 的发生和发展（贾建平，2015）。园艺康健和康复花园的运用被证实是有效的干预手段之一。

Alzheimer's disease （AD） is the most common type of dementia. Clinically, it is characterized by comprehensive dementia symptoms such as memory impairment，aphasia， apraxia, agnosia, impairment of the visual-spatial ability, executive dysfunction, and personality and behavior changes. At the beginning, the main symptoms of patients with AD include memory impairment, poor spatial orientation, susceptibility to claustrophobia, and a sleepwalking tendency, and gradually develop into cognitive dysfunction, even obnubilation, and paralysis in bed. Research indicates that AD is the most common condition leading to the loss of daily living abilities and ranks as the fifth leading cause of death in older adults. Currently, there is no complete cure for dementia; however, early engagement in cognitive training and physical exercise can delay the onset and progression of AD (Jia, 2015), such as horticultural therapy and healing garden.

AD 专类园就是专门为 AD 患者服务的康复花园，旨在帮助患者提高生活质量和幸福感，维持患者的认知和社会生活能力，以及减轻患者的精神行为症状。AD 专类园一般具有以下特征。

The AD garden is a healing garden dedicated to AD patients. It aims to help patients improve their quality of life and sense of well-being, maintain their cognition and socializing ability, and reduce their mental and behavioral symptoms. The garden generally has the following characteristics.

（1）可视性和可达性

（1）Visibility and accessibility

由室内到花园需保持视线通透，完全可见，道路明晰通达。有研究表明，室内到户外的可视性降低，会增加患者焦虑和恐惧的心理，曲折或迂回的路线会使患者有挫败感。

The sight from indoor rooms to the garden is unobstructed, and the route is clear and easy to access. Studies have shown that reducing the visibility from indoor to outdoor can increase patients' anxiety and fear, and that a tortuous or circuitous route can make patients feel frustrated.

（2）道路系统简明合理

（2）The concise and reasonable route system

针对 AD 患者记忆衰退、不辨方向的特点，AD 花园的道路多为环状回路，无死路且保持最小的选择性；主要道路宽度不小于 1.8m，适宜坐轮椅的 AD 患者并排行走。

Most of the roads in AD garden are ringlike loops, with minimal optionality and no dead ends, which caters to the characteristics of AD patients such as memory decline and inability to distinguish directions; the width of the main road is no less than 1.8 meters, suitable for AD patients in wheelchairs to move side by side.

（3）植物选择因地制宜

（3）Suitable plants for local conditions

多选用当地植物营造景观或开展园艺活动，对植物的熟悉感有助于 AD 患者建立一种归属感和与时间的关联感。

Choose more local plants to build landscapes or carry out gardening activities. Familiarity with plants helps AD patients develop a sense of belonging and connection with time.

8.2　AD 专类园设计基础理论
8.2　Foundational theories of AD gardens

　　人与环境的关系通常是由人的基本需求决定的。因此,结合健康生成论、环境心理学与情感导向的疗法,可以从最深层次去理解 AD 患者在户外环境中的需求,这是 AD 专类园设计的基础。

The basic needs of people determine the relationship between humankind and the environment. Through the combination of the salutogenesis, environmental psychology and emotion-oriented therapy, the needs of AD patients in the outdoor environment are understood at the deepest level, which is the basis for the design of AD gardens.

8.2.1　健康生成论
8.2.1　Salutogenesis

　　健康生成论是研究如何产生健康的主要理论之一,包括如何创造、增加、促进生理、心理和社交健康以达到最佳幸福感的健康策略。人的健康状况处于较低水平时,对环境的敏感程度较高,以内向活动为主导,这里的内向活动是指人被动地接受自然益处的活动。随着健康状况的好转,人对环境敏感程度降低,最初的内向活动逐渐转变为有情感参与的活动、主动参与的活动,最后转变为以外向活动为主导,使人主动从环境中获益。AD 专类园的设计要实现优化健康这一目的,从健康生成论的角度来看,需要同时将人在环境中的被动体验和主动活动都考虑在内。

Salutogenesis is one of the major theories on how to be healthy, including the strategies for how to create, increase, and promote physical, mental, and social health to achieve optimal happiness. When people's health status is at a low level, they are more sensitive to the environment and mainly engaged in introverted activities. Introverted activities refer to activities in which people passively benefit from nature. As the health status improves and the sensitivity to the environment lessens, the initial introverted activities are gradually changed to emotionally involved and actively participated ones then finally transformed into mostly extroverted ones, and as a result, people actively benefit from the environment. From the perspective of salutogenesis, it is necessary to take into account both the passive experiences and active activities of people in the environment to achieve the goal of improving health in the design of AD gardens.

8.2.2　环境压力理论
8.2.2　Environmental press theory

　　环境压力理论是 Lawton 和 Nahemow 在 1973 年开始推广的,用以解释老年人与其环境的相互作用的生态学理论。这一理论认为,老龄化是一个不断适应外部环境和内部能力及功能变化的相互作用的过程。在这个框架中，适应水平解释了老年人对环境需求的反应，反过来又促使人们做出改变，要么降低环境压力，要么提高自己的能力。该理论强调寻求个体能力与环境要求之间平衡的意义，这种平衡能增强人的幸福感。与普通人相比，AD 患者受环境因素的影响更大，随着失智症渐进性发展，患者能力水平逐渐降低，承受环境的压力逐渐增大，会产生一些不良适应行为和负面感受，导致消极后果。在 AD 花园设计中引入园艺康健是将能力—环境压力理论应用于设计的一种有效途径，通过园艺活动逐渐促进患者身体、社交和认知能力全面提升，使其感受的环境压力相应减小以致达到平衡，从而促进人的幸福感。

Lawton and Nahemow（1973）began promoting ecological theories to explain the interactions of older adults with their environments. This theory defines aging as a transactional process of continual adaptation to changes in both the external environment and internal capabilities and functioning. The environmental press theory emphasizes the meaning of seeking a balance between individual abilities and environmental requirements, and this balance can promote people's happiness. Compared with ordinary people, patients with AD are more affected by environmental factors. With the gradual development of dementia, the patients' ability level gradually decreases, and the environmental pressure they suffer gradually increases, which cause some maladaptive behaviors and negative feelings, then negative consequences will follow. The introduction of the horticultural therapy in the design of AD gardens is an effective way to apply the competence-environmental press theory to the design; the physical and social abilities of patients are gradually improved and their cognitive abilities get comprehensively boosted through gardening activities, and as a result, the environmental pressure they feel would be reduced accordingly to achieve balance, by which happiness is promoted.

8.2.3　渐进性降低压力阈值模式
8.2.3　Progressively lowered stress threshold model

该模式是指如果环境刺激超过某一个阈值，AD 患者对环境刺激做出负面反应，为避免发展为更严重的具有攻击性、破坏性的激越行为，对其所处环境做相应改变，直到负面反应减弱或消失。根据这一原理，AD 专类园在设计时应力求营造避免患者出现焦虑行为、精神症状的户外环境，例如，在尺度、栅栏、色彩、材质等方面做出调整，营造有不同刺激梯度的空间，从而减少 AD 患者因承受环境压力产生的焦虑，防止精神行为症状的发生。

The progressively lowered stress threshold model means that if the environmental stimuli exceed a certain threshold, the patients with AD will respond negatively to the environmental stimuli; the environment needs to change correspondingly until the negative response improves or disappears in order to avoid developing into more serious aggressive and destructive agitated behavior. According to this principle, the design of AD gardens should create an outdoor environment that minimizes the risk of patients' anxious behavior and mental and behavioral symptoms. An example of this is to make adjustments in size, enclosure, colors, materials, and so on, to create the space with stimuli of varying intensities, thereby reducing the patients' anxiety caused by environmental pressures and preventing the occurrence of mental and behavioral symptoms.

8.2.4　情感导向疗法
8.2.4　Emotion-oriented approaches

情感导向疗法的主要目标是将 AD 患者的注意力转移到记忆、经历和感知上，以此减少其可能产生的行为和心理问题。怀旧疗法、多感觉刺激疗法是情感导向疗法的代表疗法。在 AD 康复花园设计中引入情感导向疗法，根据 AD 患者发病阶段对应的心理年龄设计，能唤起患者记忆的引导物和场景，或设计多种环境元素建立五感刺激，可减轻 AD 患者因身体活动能力、认知能力逐渐衰弱而产生的行为和心理问题。

The main goal of emotion-oriented approaches is to divert AD patients' attention to memory, experiences, and perception, thereby reducing the possible behavioral and psychological problems. The reminiscence therapy and multisensory stimulation therapy are representative of emotion-oriented approaches. The behavioral and psychological problems caused by the gradual decline of physical activity ability and cognitive ability of patients with AD can be reduced by introducing emotion-oriented approaches in the design of AD healing gardens, setting up memory-triggering objects and scenes according to the psychological age in different disease stages of patients or designing a variety of environmental elements to stimulate the five senses.

8.3　AD 专类园案例
8.3　Cases of gardens for AD patients

8.3.1　波特兰记忆花园
8.3.1　Portland Memory Garden

坐落在美国俄勒冈州波特兰市的波特兰记忆花园是全美八个记忆公园之一，建成于 2002 年 5 月，这个花园是当地社区送给 AD 患者以及护理人员的礼物。波特兰记忆花园向整个社区开放，但其设计初衷是为 AD 患者服务，并为其看护人员提供一个休息场所。

Portland Memory Garden, established in May 2002 is one of the eight memory gardens in the U.S. The garden is open to the entire community, but was designed to meet the special needs of those with memory disorders and to provide respite for their caregivers. It's the community's gift to those with Alzheimer's disease and those who care for them.

波特兰记忆花园的设计建造充分体现了康复花园循证设计的特点，从设计之初就整合了一个强大的团队，汇集的成员来自商业、基金会、公共机构、非营利组织、高校、高中、社区、园艺俱乐部等机构等。为了设计出人人都能乐享其中的花园，志愿者付出了极大的努力。这些合作团队包括美国景观设计师协会、阿尔茨海默病协会、波特兰公园老年社会设计中心、休闲娱乐中心、莱加西医疗服务中心和波特兰州立大学城市研究和规划学院的老龄化研究所。图8-1至图8-16为美国波特兰记忆花园部分场所景观。

This garden, which is created for everyone to enjoy, is the result of a volunteer effort bringing together people of all ages from businesses, foundations, public agencies, not-profit organizations, higher education institutions, high schools, neighborhoods, garden clubs and so on. The co-developers include the American Society of Landscape Architects, Alzheimer's Association, Center of Design for an Aging Society Portland Parks & Recreation, Legacy Health System and the Institute on Aging-School of Urban Studies and Planning at Portland State University. Fig. 8-1 to Fig. 8-16 show some scenes in Portland Memory Garden.

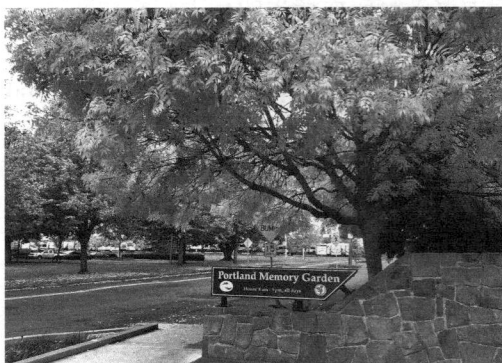

图 8-1　位于美国俄勒冈州的波特兰记忆花园(郑丽　摄)

Fig. 8-1　Portland Memory Garden in Oregon, USA (by Zheng Li)

图 8-2　入口集散小广场(郑丽　摄)

Fig. 8-2　Small square at the entrance (by Zheng Li)

图 8-3　小广场边的接待桌(郑丽　摄)

Fig. 8-3　Reception table at the small square (by Zheng Li)

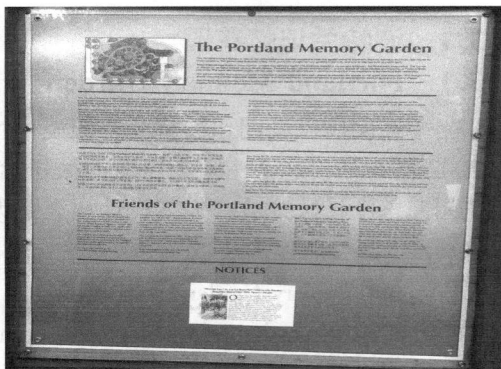

图 8-4　用五种语言介绍花园(郑丽　摄)

Fig. 8-4　Garden introduction in five languages (by Zheng Li)

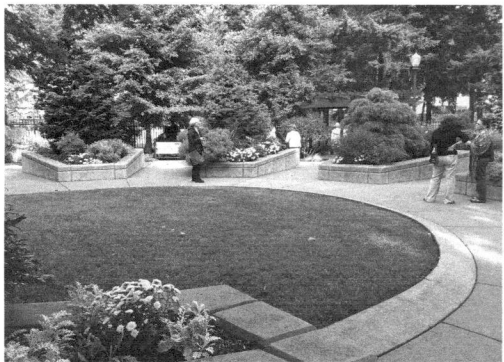

图 8-5　主花园中心草坪（郑丽　摄）

Fig. 8-5　Central lawn in main garden

（by Zheng Li）

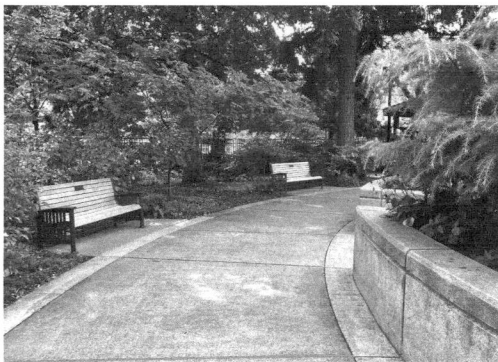

图 8-6　花园环形主路（郑丽　摄）

Fig. 8-6　Circular road in the garden

（by Zheng Li）

图 8-7　材质亲肤的座椅（郑丽　摄）

Fig. 8-7　Seats made of skin-friendly materials

（by Zheng Li）

图 8-8　丰富的植物配置（郑丽　摄）

Fig. 8-8　Rich plant configuration

（by Zheng Li）

图 8-9　特殊植物的五种语言介绍（郑丽　摄）

Fig. 8-9　Introduction to special plants in five

languages（by Zheng Li）

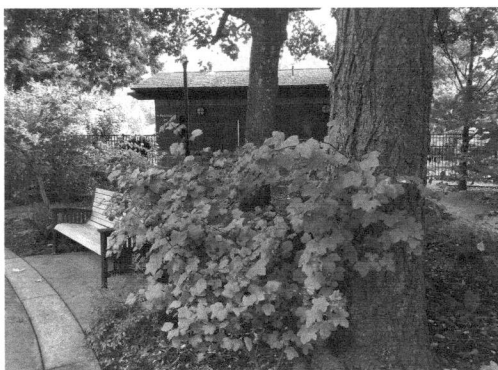

图 8-10　园艺工具房（郑丽　摄）

Fig. 8-10　Gardening toolhouse

（by Zheng Li）

图 8-11　陈列有序的工具(郑丽　摄)

Fig. 8-11　Orderly display of tools

（by Zheng Li）

图 8-12　通往菜园的砂石小路(郑丽　摄)

Fig. 8-12　Gravel path to the vegetable garden

（by Zheng Li）

图 8-13　菜园一角(郑丽　摄)

Fig. 8-13　A corner of the vegetable garden

（by Zheng Li）

图 8-14　碎木屑铺设的路面(郑丽　摄)

Fig. 8-14　Pavement paved with wood chips

（by Zheng Li）

图 8-15　质地柔和的透水砖(郑丽　摄)

Fig. 8-15　Permeable bricks with soft texture

（by Zheng Li）

图 8-16　菜园休息处(郑丽　摄)

Fig. 8-16　Rest area of vegetable garden

（by Zheng Li）

园内种植四季植物和高床栽培的花卉，用于刺激感官，唤醒对过去的记忆。花园设无障碍环形道路，可看见标志性建筑，同时外围带有栅栏。该花园免费开放。

The garden includes four seasons of plants and flowers in raised beds that have been chosen to stimulate the senses and recall past memories. The design of the garden provides barrier-free circular pathway and distinctive landmarks. The fenced garden is open from dawn to dusk all year long and is free of charge.

波特兰记忆花园的建成充分证明，只要本着建设美好社区的精神，各种公共机构、非政府组织、非营利机构以及个人志愿者的力量是无穷的。

The Portland Memory Garden is a true testimonial to what can happen when various public, NGO, and non-profit organizations, and individual volunteers come together in the spirit of creating a better community.

8.3.2　埃奇伍德—康芒斯康复花园
8.3.2　The Sedgewood Commons Healing Garden

埃奇伍德—康芒斯康复花园，位于美国缅因州法尔茅斯，因针对不同阶段的失智症患者设计花园而闻名。该花园突出了怀旧疗法的运用：根据 AD 的每个患病阶段（轻度、中度、重度）与人的心理发育阶段（成年、儿童、幼儿），相对应并产生关联而进行设计。花园收容了处于 AD 不同阶段的患者，鉴于患者认知、行为能力各不相同，为满足患者的不同需求，设计师在建筑的三个侧翼护理单元外设计了三个花园，它们的平面元素布局复杂程度逐个降低，提供的活动内容和刺激水平也各不相同，三个花园的具体情况见下文。

The Sedgewood Commons Healing Garden is located in Falmouth（Maine, USA）is a garden highlighting the use of the reminiscence therapy. Its design was made by matching and relating the levels of AD（mild, moderate, severe）to stages of mental development（adult, child, and infant）. The garden accommodates patients at different levels of AD, whose cognitive and behavioral abilities are different. To meet the different needs of those patients, three gardens were designed outside the three nursing units at the wing of the building. Their layout of the planar elements is of different complexity and the activities provided and the levels of stimulation are different. The details of the three gardens are as follows：

（1）霍桑花园
（1）The Hawthorne Garden

霍桑花园为轻微 AD 患者服务，患者相对应的心理年龄阶段为 14 周岁到成年。园中加入了能唤起这个年龄阶段记忆的旧物和园艺工具，包括新英格兰地区的传统建筑、白色尖桩栅栏、木柴堆、晒衣绳、喷壶等。

The Hawthorne Garden accommodates patients with mild AD whose corresponding mental age is from 14 years old to adulthood. Old objects and gardening tools that can evoke their memories in this age range are put into the garden, including traditional buildings in New England, white picket fences, firewood piles, clotheslines, watering cans, and so on.

(2)朗费罗花园
(2) The Longfellow Garden

朗费罗花园为中度 AD 患者服务，患者相对应的心理年龄阶段为 8～13 周岁。花园平面布局清晰明了，焦点是处于中心位置的一座希腊古翁雕塑，为患者提供了方位感。多个直角园路口让患者易于自主选择园路，同时各条道路又指向起点，营造安全感。园中最重要的元素——棚架，是患者获得安全感、方向感、围合感及探寻趣味的场所。抬高的种植床和多处喂鸟点方便乘坐轮椅的患者进行上肢活动。

The Longfellow Garden accommodates patients with moderate AD, and the corresponding mental age of those patients is from 8 to 13 years old. The layout of the garden is clear, and a sculpture of a Grecian urn in the center is the focal point, which provides a sense of direction for the patients. Multiple right-angle intersections in the garden make it easy for patients to exercise their autonomy in choosing roads in the garden, and at the same time, each road points to the starting point to create a sense of security. The most important element in the garden, the trellis, is a place for patients to gain a sense of safety, direction, and enclosure, and to explore something interesting. The raised planting beds and multiple bird feeding stations cater to patients in wheelchairs to exercise upper limbs.

(3)米莱花园
(3) The Millay Garden

米莱花园为重度 AD 患者服务，患者相对应的心理年龄阶段为婴儿到 7 周岁。米莱花园汲取了日本园林的形式和精神。为触发患者儿时溪边玩耍的记忆，设计师以老树和细长蜿蜒、枯涸的小溪作为花园的焦点。坚实的围合栅栏增强了封闭性，水平围合使患者获得心理安全感，竖向围合保证了人体安全，空间尺度合理，提高了私密性、安全性，且让患者方便接受护理和照料。此园自然要素的比重增加，使用的植物材料都是患者早年所熟知的芳香植物(此园禁种有毒植物)，包括紫丁香、金银花、月季等。

The Millay Garden accommodates patients with severe AD, whose corresponding mental age is from infancy to seven years old. The Millay Garden draws on the form and philosophy of Japanese gardens. In order to trigger patients' childhood memories of playing by a creek, the designer took old trees and a slender, winding and dry creek as the focal point of the garden. The solid fences enhance the closure; the horizontal enclosure gives patients a sense of psychological security; the vertical enclosure ensures the safety of the human body; the reasonable space size improves the privacy and safety and makes it easy for patients to be nursed and tended. The proportion of natural elements in this garden has increased, and the plant materials used are all aromatic plants (but poisonous plants are prohibited in the garden) that patients have known in their early years, including lilacs, honeysuckles, roses, and so on.

8.3.3 圣安德烈疗养院 AD 感官花园
8.3.3 AD Sensory Garden of the St. Andrew's Nursing Home

　　圣安德烈疗养院的感官花园位于新加坡的曼谷角(万国)。该疗养院是非营利的自愿福利组织圣安德烈教会医院的一部分，于 2013 年建立，为 300 名患有失智症且病情稳定的精神病状态的老人提供护理。2015 年这个总面积 1120m² 的花园在旧址花园基础上进行了重建改造。新花园于 2021 年开园，为老人、来访者以及养老院的工作人员提供了一个临时休息的空间。在设计和开发阶段，设计师与养老院管理团队紧密合作，后者包括运营人员、护士和理疗师。设计师采用了参与式设计来理解用户的需求，设计了一个以植物为主的、重点关注用户多感官体验的花园(图 8-17)。

图 8-17　新加坡万国圣安德烈疗养院的小径

Fig. 8-17　Pathway in the Sensory Garden of St. Andrew's Nursing Home (Buangkok), Singapore

　　The Sensory Garden of St. Andrew's Nursing Home is located in Buangkok Singapore. As part of the St. Andrew's Mission Hospital, a non-profitable, voluntary welfare organization, the nursing home was established in 2013 and has provided care for 300 residents with dementia and stable psychiatric conditions. The 1120 square meters garden is a redevelopment of the original back in 2015. Launched in 2021, the new garden provides a space for respite for both residents, visiting family members, as well as the nursing home staff. During the design and development process, the designers worked closely with the nursing home management team, including operational staff, nurses and physiotherapists. A participatory design approach was used to understand the needs and requirements of the users. The result is a garden focused on multi-sensory experiences for the users with plants(Fig. 8-17).

　　结合注意力恢复理论和压力缓解理论的环境心理学原则，感官花园设计成六个区域：五个感官区和一个怀旧区。当患者在不同的区域活动时，每个区域都为他们提供不同的体验。

　　Based on principles of environmental psychology from Attention Restoration Theory and Stress Reduction Theory, the garden is choreographed into six zones：five sense zones and one reminiscence zone, each providing different experiences to the users as they move around.

(1)第 1 区(嗅觉)

(1)Zone 1-Smell

在失智症的早期阶段，短期记忆等认知功能丧失，而长期记忆通常完好无损。研究表明，气味是最强烈的感觉之一，可以触发患者的情绪和长期记忆。因此，感官花园在一开始就引入熟悉的芳香类植物。患者在入口处就能闻到纽子花属植物的清香，而栀子花和水梅则使他们能够联想和这种气味有关的记忆。

In the early stage of dementia while cognitive functions such as short-term memory lapses, long-term memory often stays intact. And studies have shown that smell is one of the strongest senses that can trigger emotions and long-term memories. Hence, common fragrant plants are introduced at the beginning. The faint scent of the Bread Flower(*Vallaris*) greets the participants at the entrance, while others like the Cape Jasmine (*Gardenia jasminoides*) and the Water Plum (*Wrightia religiosa*) enable them to recall their past memories relating to the smell.

(2)第 2 区(听觉)

(2)Zone 2-Hearing

本区引入了各种各样的草类植物，能够在风中沙沙作响，如黄扇鸢尾和竹叶兰。此外，吊裙草干燥的种子荚也可以发出咔嗒声，使患者在与植物互动时产生兴趣。木质风铃和大型竹制乐器也被引入园内，以在花园的道路上创造出一些互动的声音体验(图 8-18)。

图 8-18　新加坡万国圣安德烈疗养院感官花园的听觉区和怀旧区之间的通道

Fig. 8-18　Pathway between the hear zone and reminiscence zone at the Sensory Garden of St. Andrew's Nursing Home (Buangkok), Singapore

Here a variety of herbaceous plants that trigger a rustling sound in the wind, such as the Yellow Walking Iris (*Trimezia martinicensis*) and Bamboo Orchid (*Arundina graminifolia*) are introduced. In addition, the dry seed pods of the Rattleweed Plant (*Crotalaria retusa*) can also be shook to create a rattling sound, arousing interest of the users as they interact with the plant. Wooden wind chimes and large bamboo musical instruments are also introduced for some interactive sound experiences along the garden path(Fig. 8-18).

(3)第 3 区(怀旧)
(3) Zone 3-Reminiscence

本区展示了具有文化意义的植物和过去常种植的植物,包括莲花、石榴、芙蓉花、五彩苏、香露兜、蝴蝶花豆,这些都是过去随处可见的植物。此外,花园亦增设了一个拉茶小吃摊位*(木制手推车),让老年用户增加就餐选择。该区域还可以作为活动空间进行团体活动,包括可在拉茶小吃店摊位上提供聚餐。

Plants of cultural significance and those that were commonly grown in the old days are showcased here, such as the Lotus (*Nelumbo nucifera*), Pomegranate (*Punica granatum*), Hibiscus (*Hibiscus rosa-sinensis*), Common Coleus (*Coleus scutellarioides*), Pandan (*Pandanus amaryllifolius*) and Butterfly Pea (*Clitoria ternatea*). A Sarabat stall (wooden pushcart)* was also introduced to let the elderly users relive the old days of eating at roadside food stalls where street foods were sold. This zone also doubles as a site for group activities, including meal gatherings where food can be served on the Sarabat stall.

(4)第 4 区(味觉)
(4) Zone 4-Taste

紧邻怀旧区的是味觉区,这里种植着蔬菜、果树、药草和香料等可食用植物。菜地让工作人员和患者在此共同参与园艺活动,这是另一种锻炼身体技能(包括精细和总体运动技能)的良好训练场地。可移动的高床花槽方便轮椅使用者与植物互动,即以舒适的姿势播种和收获蔬菜,不需要弯腰或从轮椅上站起来。即使是那些无法参与体力活动的患者也可以简单地在这一区域散步,可以捣碎、嗅闻和品尝草本植物的叶子,如罗勒、到手香和北艾(图 8-19)。

Next to the reminiscence zone is the taste zone where edible plants such as vegetables, fruit trees, herb and spices are planted. The vegetable plots allow both the staff and patients to get physically involved in gardening activities, another good exercise to practice their physical skills, including fine and gross motor skills. The raised mobile planter also allows wheelchair-bound users to interact with the plants i. e. , sow of seeds and harvesting of vegetables in a comfortable position without the need to bend or get up from the wheelchair. Even for those who are unable to get involved in the physical activity, they can simply walk

---

* 拉茶小吃:是东南亚一种传统的奶茶,通常用木制小推车搭配零食售卖。

Sarabat stall: a traditional milk tea from Southeast Asia, usually sold from wooden pushcarts along with snacks.

around this area, crushing, smelling and tasting the leaves of herbs, such as the Common Basil (*Ocimum basilicum*), Indian borage (*Coleus amboinicus*), as well as the Common Mugwort (*Artemisia vulgaris*) (Fig. 8-19).

(5)第 5 区(触觉)

(5)Zone 5-Touch

紧邻味觉区的是触觉区，该区植物的叶子和茎有不同触感，在质感柔软的爆仗竹、质地光滑的龙船花叶片，以及五爪木复伞形花序的有趣纹理上都可以感受到不同触觉的差异(图 8-20、图 8-21)。

Right beside the taste zone is the touch zone which displays plants with leaves and stems of various tactile qualities, for example the soft texture of the Firecracker Plant (*Russelia equisetiformis*), the waxy leaves of the Ixora (*Ixora chinensis*), and the interesting texture of the flower clusters of the Green Aralia (*Osmoxylon lineare*) (Fig. 8-20, Fig. 8-21).

图 8-19　新加坡万国圣安德烈疗养院感官花园的味觉区和触觉区之间的通道

Fig. 8-19　Pathway between the taste and touch zones at the Sensory Garden of St. Andrew's Nursing Home (Buangkok), Singapore

图 8-20　新加坡万国圣安德烈疗养院感官花园的触觉区与视觉区域之间的通道

Fig. 8 20　Pathway between the touch and sight zones at the Sensory Garden of St. Andrew's Nursing Home (Buangkok), Singapore

图 8-21   新加坡万国圣安德烈疗养院感官花园的触觉区草坪景观

Fig. 8-21   **The lawn area in the touch zone at the Sensory Garden of St. Andrew's Nursing Home（Buangkok），Singapore**

（6）第 6 区（视觉）
（6）Zone 6-Sight

感官花园体验的最后一站是视觉区，展示的是提供大量趣味视觉的植物类型，如响尾蛇竹芋、苏里南朱缨花和云桂叶下珠。在这里，其他感官特征亦被综合呈现，如有花朵的香味和水景的潺潺声，明亮温暖的色彩增强了视觉感受，还有在草坪区域上行走的触觉体验。

The garden experience ends with the Sight Zone where the plant types on display allow for a lot of visual interest, such as the Rattlesnake Plant（*Calathea crotalifera*），the Red Powderpuff Plant（*Calliandra surinamensis*）and the Tropical Leaf-flower（*Phyllanthus pulcher*）. Here, other sensory features are reintroduced: the aroma of flowers, the trickling sound of water features, the strong visual impact from bright and warm colors, and the tactile experience of walking on the lawn area.

随着路径的循环回到花园的起点，患者可以继续探索花园中之前可能错过的地方。回到同一方向的环路可以防止他们迷路或感到困惑。工作人员和护理人员也可以确保患者的安全和行踪。在花园中，沿着小路每隔 5m 就有一条长凳。每隔一定距离设置的长凳能够鼓励老年人继续探索花园，使他们确信如果走累了就可以休息。长凳面向不同的方向，提供不同的风景，长凳被置于可刺激感官的植物之丰富的色彩、纹理和气味中。

As the path loops back to the start of the garden, the patients can continue to explore the parts of the garden that they might have missed out earlier. A loop back to the same direction also prevent them from getting lost or confused. Staff and caregivers can also be assured of their safety and whereabouts. In the garden, a bench is placed every 5 metres along the path. The benches at regular intervals encourage the elderly participants to continue to their exploration in the garden, reassuring them that they will be able to rest should they are tired from walking. The benches are also placed to face different directions to provide a variety of views, and also amongst the profusion of color, texture and scents of the plants which stimulate the senses.

总之，感官花园为久病在床的患者提供了一个放松之地。即使是那些无法下床的患者，工作人员也可以把他们推到花园周围，因为小路足够宽，可以推行病床。哪怕不去花园，只需望着窗外花园中的绿色植物，在一定程度上就能减轻疲劳。花园活动后的总结回顾受到了工作人员和患者的一致好评。这种使用后评估(POE)可确保花园特色是否很好地得以利用。与位于公园中的康复花园类似，万国圣安德烈疗养院的老年失智症花园，也允许患者从事园艺和其他园艺康健活动，这大大加快了患者的康复进程。

Overall, the garden provided relief and refuge for the residents who usually spend most of their hours lying in bed. Even for those who are unable to get out of their bed, caregivers can bring them to the garden as the paths are wide enough for the hospital beds to be pushed around. Even if they are not brought there just by looking out of the windows towards the greenery in the garden, their fatigue can get relieved to some degree. Post Implementation Review (PIR) was warmly received by the staff and residents, and a Post Occupancy Evaluation (POE) will also be conducted to determine if the garden features are well utilized. Similar to the healing gardens in public parks, the dementia garden at St. Andrew Nursing Home also allows residents to engage in gardening and other therapeutic horticulture activities. This greatly enhances the healing and rehabilitative process of the residents.

思考题

请简述 AD 专类园的特点。

Question

Please describe the characteristics of the AD gardens.

第 *9* 章

儿童康复花园常见类型

Chapter 9　Common Types of Healing Gardens for Children

本章介绍的儿童康复花园主要包括儿童医院花园，以及针对特殊疾病，如孤独症患者设计的专业性康复花园两大类型。儿童因其年龄的特殊性，对自然充满兴趣，喜欢亲近自然、探索空间，也容易通过接触自然元素以及相关活动，从病痛的折磨中得到放松，从而触发多种积极的生理反应，如降低血压、缓解肌肉紧张、降低皮肤的电传导率等。

The healing gardens for children introduced in this chapter include gardens of children's hospitals and the professional healing gardens designed for patients with special types of diseases such as autism. Because of their age, children are interested in nature. They like to get close to nature and explore space. Despite the torment of illness and pain, they get relaxed easily through contact with natural elements and related activities, and thereby a variety of positive physiological reactions are triggered, for example, a lowered blood pressure, alleviated muscle tension, lowered electrical conductivity of the skin, and so on.

9.1　医院内儿童康复花园

9.1　The healing garden for children in a hospital environment

医院内的儿童康复花园与普通医疗环境康复花园有着很多相似性，基本设计流程和设计方法一致。所不同的是，由于使用群体以生病儿童、陪护患儿的父母以及兄弟姐妹或者其他小朋友等为主，需要从满足各部分群体的需求考虑。如需要考虑设计以下几种不同使用功能的花园：健康孩子可以积极活动的场地，针对患者儿童及父母使用的被动式"治疗空间"、冥想花园，给医务工作人员使用的小花园等，图9-1至图9-37列举了美国Legacy儿童医院的部分场景。

The children's healing garden in a hospital environment has many similarities with the healing gardens in a general medical environment, and their basic design processes and methods are the same. The difference is that as the user groups are mainly sick children, parents who are accompanying them, siblings or other children and so on, it is necessary to consider and meet the different needs of those groups. It is necessary to consider the design of some types of gardens for different uses: a venue where healthy children can engage in activities, the passive "healing space" for sick children and their parents, small gardens for staff use, and so on. Fig. 9-1 to Fig. 9-37 showcase some scenes from Legacy Children's Hospital in the United States.

图 9-1 Legacy 儿童健康中心（郑丽 摄）
Fig. 9-1 Legacy Children's health center
（by Zheng Li）

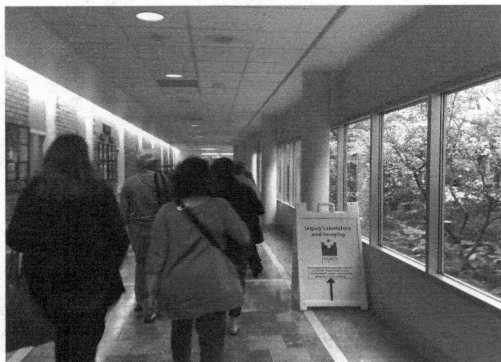

图 9-2 带玻璃墙的走廊（郑丽 摄）
Fig. 9-2 The corridor with a glass wall
（by Zheng Li）

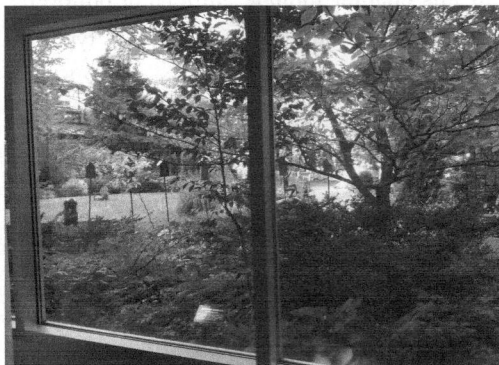

图 9-3 走廊一侧的花园窗景（郑丽 摄）
Fig. 9-3 The garden window scene on one side
of the corridor（by Zheng Li）

图 9-4 充满童趣的窗景（郑丽 摄）
Fig. 9-4 Childlike window view
（by Zheng Li）

图 9-5　充满童趣的墙画（郑丽　摄）
Fig. 9-5　**A wall painting full of childlike interest**（by Zheng Li）

图 9-6　空中花园入口（郑丽　摄）
Fig. 9-6　**The entrance to the sky garden**（by Zheng Li）

图 9-7　正对入口的花坛（郑丽　摄）
Fig. 9-7　**The flower bed facing the entrance**（by Zheng Li）

图 9-8　通向露台的无障碍坡道（郑丽　摄）
Fig. 9-8　**The wheelchair ramp to the terrace**（by Zheng Li）

图 9-9　花园上方的露台（郑丽　摄）
Fig. 9-9　**The terrace above the garden**（by Zheng Li）

图 9-10　俯瞰花园（郑丽　摄）
Fig. 9-10　**Overlooking the garden**（by Zheng Li）

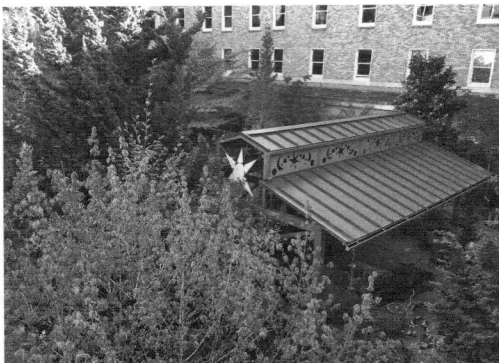

图 9-11　俯瞰凉亭(郑丽　摄)

Fig. 9-11　Overlooking the pavilion
（by Zheng Li）

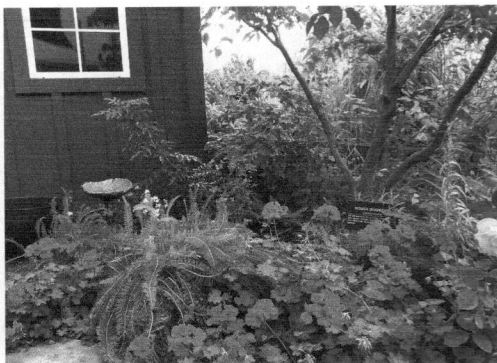

图 9-12　丰富的植物配置(郑丽　摄)

Fig. 9-12　Rich plant species
（by Zheng Li）

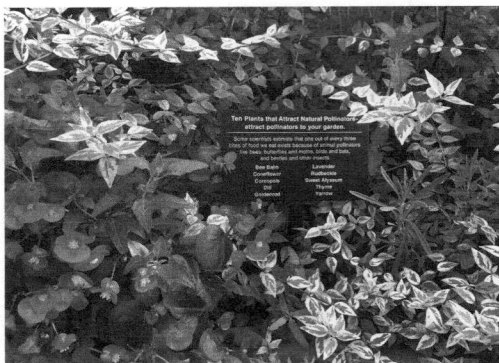

图 9-13　小科普：吸引传粉者的十大虫媒花
（郑丽　摄）

Fig. 9-13　Science popularization—ten plants that attract natural pollinators（by Zheng Li）

图 9-14　舒适的休憩设施(郑丽　摄)

Fig. 9-14　Comfortable sitting facilities
（by Zheng Li）

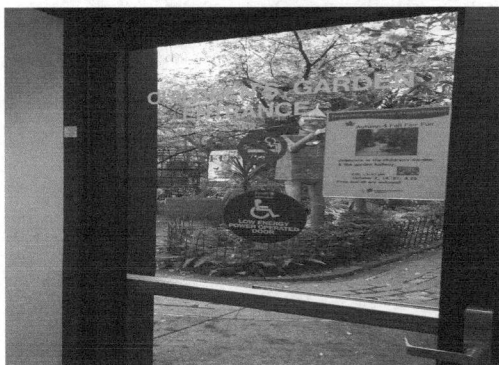

图 9-15　儿童花园入口(郑丽　摄)

Fig. 9-15　The entrance to the children's garden（by Zheng Li）

图 9-16　花园入口处的凉亭(郑丽　摄)

Fig. 9-16　The pavilion at the entrance to the garden（by Zheng Li）

图 9-17　小科普：食物长成记（郑丽　摄）

Fig. 9-17　Science popularization—food growth
（by Zheng Li）

图 9-18　活泼的标识牌（郑丽　摄）

Fig. 9-18　A lively sign（by Zheng Li）

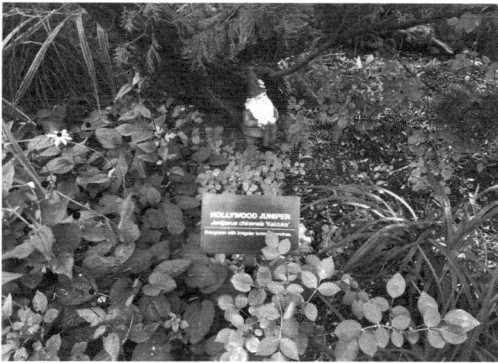

图 9-19　花园里的童话（郑丽　摄）

Fig. 9-19　Fairy tales in the garden
（by Zheng Li）

图 9-20　别致的铭刻墙（郑丽　摄）

Fig. 9-20　A chic engraved wall
（by Zheng Li）

图 9-21　花园铭句（郑丽　摄）

Fig. 9-21　The garden inscription
（by Zheng Li）

图 9-22　铜人水管（郑丽　摄）

Fig. 9-22　The copper portrait water pipe
（by Zheng Li）

图 9-23　无处不在的座椅（郑丽　摄）

Fig. 9-23　Seats available everywhere
（by Zheng Li）

图 9-24　精美的鸟屋（郑丽　摄）

Fig. 9-24　A beautiful birdhouse
（by Zheng Li）

图 9-25　鸟屋（郑丽　摄）

Fig. 9-25　Birdhouses（by Zheng Li）

图 9-26　草坪（郑丽　摄）

Fig. 9-26　The lawn（by Zheng Li）

图 9-27　五彩牛（郑丽　摄）

Fig. 9-27　The colorful bull（by Zheng Li）

图 9-28　铁皮人（郑丽　摄）

Fig. 9-28　The tin man（by Zheng Li）

图 9-29　有趣的角落（郑丽　摄）
Fig. 9-29　An interesting corner
（by Zheng Li）

图 9-30　鸟语花香处（郑丽　摄）
Fig. 9-30　The birds' twitter and fragrance of flowers（by Zheng Li）

图 9-31　欢声笑语（郑丽　摄）
Fig. 9-31　Cheers and laughter
（by Zheng Li）

图 9-32　园艺工具房（郑丽　摄）
Fig. 9-32　The horticultural tool room
（by Zheng Li）

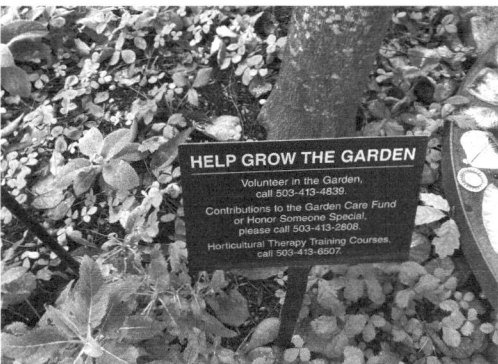

图 9-33　管理维护提示牌（郑丽　摄）
Fig. 9-33　A garden sign
（by Zheng Li）

图 9-34　波浪草坪（郑丽　摄）
Fig. 9-34　The wavy lawn（by Zheng Li）

图 9-35　医护休息区（郑丽　摄）

Fig. 9-35　**The medical personnel rest area**

（by Zheng Li）

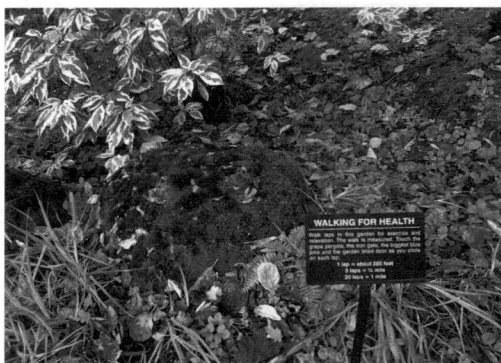

图 9-36　花园健身提示（郑丽　摄）

Fig. 9-36　**A garden sign to encourage walking**

（by Zheng Li）

图 9-37　蜂鸟和蝴蝶的花园（郑丽　摄）

Fig. 9-37　**The garden of hummingbirds and butterflies**（by Zheng Li）

归纳起来，医院内儿童康复花园必须注意的设计要点如下：

In summary, the key design points for the children's healing garden in hospitals are as follows.

①分散设置各种活动设施。为那些敏感的、害羞的儿童，或者残疾儿童提供适合他们参与活动的设施，他们通常不太愿意到设施集中区域去活动。同时，设施的安排也要考虑不同年龄段孩子的需求。

①Arrange various facilities for activities in a decentralized way. For children who are sensitive, shy, or have some disabilities, it is necessary to provide some facilities suitable for their participation in activities, as they are usually not willing to use the facilities that are used intensively. At the same time, the needs of children of different ages should also be considered in the arrangement of facilities.

②为患儿提供有趣且能分散注意力的景观，如奇异的或者令人惊喜的元素，这些患儿可能会被带到花园里，他们坐在轮椅上，或者带着输液架，甚至躺在担架上。

② Provide interesting and distracting landscapes for sick children, such as strange or surprising elements. Those children, sitting in a wheelchair, having an infusion stand with them, or even lying on a stretcher, may be taken to the garden.

③为花园设计一个明确的、坚实的边界。患儿的父母可能会带着其他孩子一起来医院看护，要确保健康的孩子在花园中奔跑或者探险时不会走丢。

③Design a clear and solid boundary for the garden. Parents of sick children may bring other children to the hospital when tending the sick ones, so it is necessary to ensure that healthy children will not get lost when running or exploring in the garden.

④为来访的家人以及医护人员提供充足而舒适的座椅，以及能够带动多感官体验的景观。座椅的位置选择很关键，座椅上休息的家长能够很方便地看到孩子活动的空间，确保孩子的安全。提供多样化的座椅，满足不同使用者的需求，例如，适合儿童尺度的可活动的座椅、适合家长看护休息的座椅、适合员工休息的私密性座椅等。

④Provide adequate and comfortable seats and the landscapes that can bring multi-sensory experiences for visiting family members and medical professionals. The position of seats is very important. Parents resting on the seats should easily see the space where the children have activities to ensure their safety. A variety of seats should be provided to meet the needs of different users, for example, movable seats for children, seats for parental care and rest, and private seats for the staff to rest.

⑤道路设计要考虑能够激发患儿游玩的兴趣，让他们愿意来花园中活动，如一条通往秘密隐藏地的道路，或者能够将患儿带入多样空间体验的道路等。可以考虑引入地形元素，丰富景观体验。

⑤The road design should arouse children's interest in play, and make them willing to have activities in the garden, for example, we can design a road leading to a secret hiding place, or a road that can bring a variety of space experiences to children. The introduction of topographical elements can be considered to enrich the landscape experiences.

⑥尽早咨询康复设计师，需要在哪些方面与康复治疗性景观结合，比如扶手高度、不同边界处理、不同的路径表面、不同的植物类型等。

⑥Consult the rehabilitation designer as soon as possible on what aspects need to be combined with the healing and therapeutic landscape, such as the height of the handrail, treatments of different boundaries, different pathway surfaces and plant types, and so on.

⑦对员工进行培训，让其了解花园设置的目的、功能、位置，让医护人员能够把治疗与花园的使用结合起来。

⑦Train employees to make them understand the purpose, function, and location of the garden, so that medical professionals can combine therapies with the use of the garden.

9.2　孤独症康复花园

9.2　Healing gardens for children with autism

孤独症，又称自闭症，是一种发育障碍类疾病，《美国精神疾病诊断及统计手册》将孤独症的核心症状归结为社交障碍，沟通困难和有限的、重复和刻板的行为，具有这些症状的疾病总称为孤独症谱系障碍（ASD），其中较严重的状况称为孤独症或经典（典型）孤独症谱系障碍。ASD 在人类各种族、各年龄段和各经济阶层中都会出现，其特征和严重程度各异，病程可持续一生，难以逆转。孤独症在儿童阶段尤为突出，且患病率逐年升高，给家庭和社会带来巨大的负担。据不同国家的统计表明，孤独症在总人口的患病比例可达2%甚至更高。在美国，每个孤独症儿童一生的护理费用超过320万美元，全国每年的花费超过350亿美元。在我国，据2022年两会发布的关于孤独症儿童数据显示，中国孤独症发病率为0.7%，14岁以下的孤独症儿童约有200万人。

Autism, also known as Autism Spectrum Disorder (ASD), is a developmental disorder. *The Diagnostic and Statistical Manual of Mental Disorders* identifies the core symptoms of autism as social communication impairments, communication difficulties and restricted, repetitive and stereotyped behaviors. Diseases with these symptoms are collectively called ASD, and the more serious conditions of the diseases are called autism or classic (typical) ASD. ASD occurs in all races and economic classes of humankind, and may affect people of any age, varying in characteristics and severity; the course of the disease can last a lifetime and is difficult to reverse. Autism is an important disease that causes mental disability in children with its prevalence increasing year by year, having caused a huge economic burden on families and society. According to statistics from different countries, the prevalence of autism in the total population has reached two percent or even higher. In the United States, the lifetime care cost of each child with autism exceeds USD 3.2 million, and the national annual expenditure exceeds USD 35 billion. In China, according to data on children with autism released during the Two Sessions in 2022, the prevalence of autism in the country is 0.7%, with over 2 million children under the age of 14 affected by the condition.

研究表明，孤独症是由环境因素、生物因素和遗传因素共同引起的，其中环境因素越来越引起研究者的关注。迄今为止，人们对孤独症的致病机理尚未完全了解，也还没有切实有效的针对核心症状的治疗方法。目前，治疗孤独症主要采用行为干预和药物干预，以及特殊教育法、生物医学干预法和心理干预法等方法，其中常用的是使用高度结构化和密集的技巧性训练的行为干预法，来帮助儿童发展社会和语言技能。包括应用行为分析、感觉统合训练、结构化教育等手段。这些治疗方式多数是在室内进行，存在治疗需时长、频率高、效果不明显、易复发等缺陷。

Studies have shown that autism is caused by environmental, biological, and genetic factors, and environmental factors have attracted more and more attention from researchers. However, so far, people have not fully figured out the pathogenesis of autism, and there is no practical and effective treatment for the core symptoms. The current treatment for autism mainly uses behavioral intervention and drug intervention, as well as special education and biomedical and psychological intervention methods. Among them, the behavioral intervention methods that use highly structured and intensive technical training to help children develop social and language skills are most used, including the applied behavior analysis, sensory integration training, structured education and other means. Most of these treatment methods are performed indoors with shortcomings like long treatment time, high frequency, an insignificant effect, and possible disease relapses.

现代医学研究表明，景观对大脑皮质和心理状态有良好的调节作用，参照孤独症治疗基本原理，借助自然和人工景观对孤独症进行干预治疗存在理论可能性。

Modern medical research has shown that landscapes have a good regulatory effect on the cerebral cortex and mental state. Based on the basic principles of autism treatment, it is theoretically possible to intervene and treat autism with the help of natural and artificial landscapes.

9.2.1　孤独症群体的基本诉求
9.2.1　The basic demands of the autistic group

孤独症群体具有社交障碍、交流障碍和刻板重复行为等症状。首先，患者在不同时期均存在一定程度社交障碍问题，通常不擅长与人交流，缺乏情感倾诉的能力，很难与他人建立较为亲密的关系。其次，是患者交流上的障碍，表现为语言理解力的缺失、语言发育减缓、语言形式、内容异常和语言运用能力的异常。患者无法用言语表达自己的情感需求，不能和同龄人进行沟通，导致与正常人的关系越来越疏远。再次，是患者兴趣狭窄和行为上的刻板重复，他们往往会专注在一些形状特别的物品上，机械地重复某一特定动作，表现出类似强迫症的行为方式。最后，患者还会存在其他状况，如精神发育迟缓或具有某些卓越突出的才能等。

Autistic groups have symptoms such as impaired social interactions and communication, and stereotyped and repetitive behaviors. First, they have problems in social interactions in different periods. They are usually not good at communicating with others, lacking the ability to express emotions, and it is difficult for them to establish a relatively intimate relationship with others. Second, they have problems in communication, which are manifested in the lack of the language comprehension ability, the slowdown of language development, and the abnormality of language forms and content and of the ability to use language. They cannot express their emotional needs in words and communicate with their peers, resulting in being increasingly isolated from normal people. Third, they usually have narrow interests and stereotyped, repetitive behaviors. They tend to focus on some special-shaped objects, repeat a specific action mechanically, and show behaviors similar to those of the obsessive-compulsive disorder. Fourth, there are also some other conditions, such as mental retardation or some outstanding talents.

根据孤独症群体的生理、心理发育特点（包括他们的感知觉与动作发展的特点），孤独症儿童康复的基本诉求包括三个层面的内容，第一个层面是身体机能发展，包括模仿、感知觉、大小肌肉的训练；第二个层面是自理技能发展，包括穿衣、进食、洗漱等；第三个层面是社交能力发展，包括情绪情感控制能力、认知理解和表达的能力和行为问题。研究表明，对于孤独症群体的治疗干预越早，康复效果越明显，故而目前研究多集中在孤独症儿童患者的治疗和环境（康复花园）建设上，对成人患者的关注较少。基于孤独症核心症状，针对成人患者的康复环境设计要求主要表现在可预测、易于理解和减少过度刺激。

According to the characteristics of the physiological and psychological development of the autistic group (including the characteristics of their perception and motor development), the basic demands during the rehabilitation of autistic children include three aspects. The first aspect is the development of somatic functions, including imitation, perception, and training of big and small muscles; the second is the development of self-care skills, including dressing, eating, washing one's face, rinsing one's mouth and so on; the third is the development of social skills, including the ability to control emotions and feelings, the capacity to perceive, understand, and express oneself, as well as behavior problems. Studies have shown that the earlier the treatment and intervention are provided for the autistic group, the more obvious the rehabilitation effect is. Therefore, the current research usually focuses on the treatment of children with autism and the environment (healing gardens) construction, and less attention is paid to adults with autism. Based on the understanding of the core symptoms, the requirements of adults with autism for the environment are mainly focused on predictability and simplicity to avoid destructive emotional stimulation.

9.2.2 孤独症康复花园的设计要求
9.2.2 Design requirements of healing gardens for people with autism

(1)场所的安全性
(1)The safety of premises

孤独症患者沉浸在自己的世界里，对环境危险往往缺乏警醒，因此在景观设计时，如设计植物、铺装、休息设施等，安全性必须放在首位。

Immersed in their own world, autistic patients usually lack awareness of environmental hazards. Therefore, safety is a top priority for the design of environmental landscapes. Safety should be paid attention to in the design of plants, pavements, rest facilities, and so on.

(2)营造开展"五感体验"自然环境空间
(2)Create the natural environment and space for "experiences of five senses"

康复花园的自然要素是孤独症患者进行五感体验的主要媒介，孤独症康复花园要利用温暖阳光、新鲜空气、生机勃勃的植物、潺潺流水等元素造就自然环境，以使患者心情得到放松和愉悦。同时，利用环境对视觉、嗅觉、听觉、触觉和味觉五个感官的刺激，激发患者思维。

The natural elements of healing gardens are the main medium for patients with ASD to have experiences of five senses. The ASD garden should use warm sunlight, fresh air, vibrant plants, gurgling water and other elements to form a natural environment to relax and delight patients mood, and bring to them the simulation from the five senses of sight, smell, hearing, touch and taste to inspire thinking.

(3)空间简洁明了，活动针对性强
(3)**The space should be concise and clear, and its activities highly targeted**

孤独症患者对外在环境感知较差，且复杂的环境不利于其情绪的平静，故孤独症

康复花园强调空间设计目的要明确，布局简单，空间边界清晰，路线明确，场所设施突出活动特点。如感统器材训练的场所，器械不可太多太杂。

Autistic patients have poor perception of the external environment, and a complex environment is not conducive to their emotional calm, so the ASD garden emphasizes that the purpose of its space design should be clear, the layout simple, the space boundary and the route clear, and that the venues and facilities should highlight the characteristics of activities, for example, in the venue for the equipment-assisted sensory integration training, there should not be too much or too miscellaneous equipment.

（4）场地具有可改造性
（4）The venue should be retrofittable

孤独症的治疗是一个循序渐进的系统工程，需要设计成多种不同阶段的活动方案。作为行为干预治疗的室外空间，服务于孤独症患者的康复花园，应具备根据不同活动需求而进行空间调整优化的可延展性。

The treatment of autism is a step-by-step system designed into activity programs of different stages. As the outdoor space for the behavioral intervention therapy, the healing garden provided for autistic patients should be adjustable according to different activity demands, and the space optimization should be highly malleable.

9.2.3　康复景观干预模式
9.2.3　The healing landscape intervention model

康复景观干预平台构筑了儿童孤独症辅助疗法体系（图 9-38），每个阶段都有既定的康复方法和康复目标，阶段之间是相互关联和促进的关系，最终通过系统干预获得持续的发展。具体来讲，第一阶段"五感体验"，即自然助益的过程，前提是处于康复花园之中并有充裕的体验时间；第二阶段"激发思维"，由感官刺激引发思维活动，由于孤独症儿童兴趣狭窄，很少自主与外界环境发生信息交换，因此通过刺激引起注意分散是康复花园干预的关键环节；第三阶段是通过主动和被动方式的康复活动促进机体的发展。

The healing landscape intervention platform has constructed an adjuvant therapy system for children with autism (Fig. 9-38), with established healing methods and goals for each stage. The stages are interrelated and mutually promoted and continuous development can be achieved through systematic intervention in the end. Specifically, the first stage of "experiences of five senses" is a process of benefiting from nature, whose premise is having been in the healing garden with ample experience time; in the second stage of "stimulating thinking", sensory stimulation triggers thinking activities, As children with autism have narrow interests and rarely exchange information autonomously with the external environment, distracting attention through stimulation is a key link in the intervention of the healing garden; the third stage is the promotion of the somatic development by active and passive rehabilitation activities.

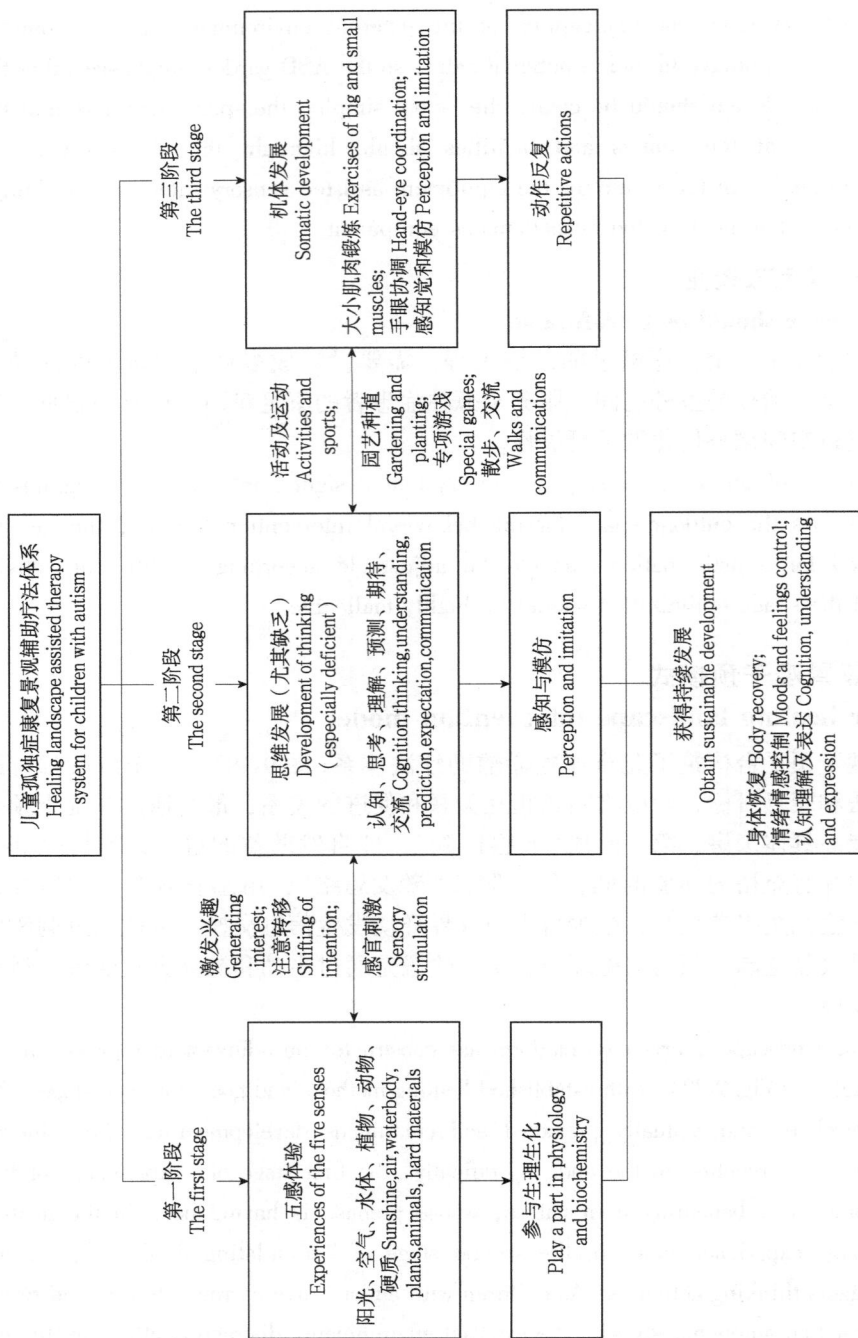

图9-38　孤独症医疗康复景观体系

Fig. 9-38　The medical rehabilitation system of healing landscapes for ASD

在孤独症患者的康复介入工作中，要有针对性地进行园艺活动设计，以促成其整体感统的协调。

In the intervention work for the rehabilitation of these patients, we must carry out a targeted gardening design to promote their overall sensory integration.

9.2.4 孤独症康复花园设计要点
9.2.4 Design key points of healing garden for people with autism

（1）园地的地形选择
（1）The terrain selection for gardens

孤独症患者的性格特点决定了孤独症康复花园不应是开阔型，或便于人群汇聚的地形，那样不利于患者情绪康复。而应采用缓坡地形，且用植物形成障景，不可有过于复杂的台阶，通道上应该设置扶手，以方便行动不便儿童的行走。

The personality characteristics of children with autism determine that the garden for persons with ASD must not be an open one, and any design that gets people together is not conducive to emotional recovery. A gently sloping terrain should be adopted; plants should be used to form obstructive scenery; there should not be overly complicated footsteps; and handrails should be set on the passage to facilitate the walking of children who have difficulty in moving.

（2）水景
（2）Waterscapes

有些孤独症儿童对水有明显的恐惧，会造成情绪失控，应尽量避免在花园内设置大型水面设施。

Some children with autism have a clear fear of water, which can cause loss of emotional control. Large-scale facilities on the water should be avoided in the park.

（3）活动区域
（3）The activity area

组织孤独症患者的活动，人数一般3~5人，这就要求活动区域的设计应有所控制。可采用地形、或高低不同的植物进行分割，形成相对私密的小型空间；但空间内的活动情况应该便于监护者从外面进行观察。如采用石头取代植物的迷宫设计，不仅安全，还能训练感官协调（图9-39）。

The activities participated in by less than 3 - 5 children would be suitable for children with autism. This requires that

图9-39 石子迷宫（郑丽 摄）
Fig. 9-39 Stone labyrinth（by Zheng Li）

the activity area in the design should be controlled. It can be divided by terrains or plants of different heights to form relatively private small space, but parents should be allowed to observe the activities in the space from outside. For example, the maze design using stones instead of plants not only is safe, but also can train sensory coordination (Fig. 9-39).

（4）设施

（4）Facilities

在孤独症儿童的活动空间内，不应该出现可移动的物件，尤其是小型的设施零件类；因为患儿可能会把物品放进嘴里，如果零件过小，可能会导致事故发生；大型可移动零件或物品也可能被患儿当作武器使用。

In the activity space of autistic children, there should be no movable objects, especially small facility parts. As those children may put objects in their mouths, too small parts may cause accidents; and large movable parts or objects might be acquired by children and then used as weapons.

（5）植物选择

（5）The plant selection

不种植任何带刺、具有刺激性、茸毛类的植物；任何具有特殊不适气味的植物也不适合栽种；花季会产生大量花粉的植物以及具有毒性或腐蚀性汁液的植物也禁用。

Any thorny, pungent, fluffy plants are not allowed; any plants with obvious smell are not suitable; plants that produce a lot of pollen in the flowering season and ones that are toxic or have corrosive sap are forbidden.

孤独症儿童园艺康健的主要目标，是改善其触觉钝化，搭建植物沟通桥梁，可以进行如下营造：

The main goal of the horticultural therapy for autistic children should be to enhance their tactile inactivation and build a communication bridge with the help of plants. The following settings can be made：

①运用彩叶植物围成小型封闭性环境，可帮助患儿在情绪发生问题时迅速安静下来。

①Use colorful herbaceous plants to form a small enclosed environment, where children can quickly calm down, and it would be a good activity area when they have emotional problems.

②应用蝴蝶花、雏菊、一串红、薄荷类植物，设计成可以提供自由采摘的活动空间，从而进行采摘作业或进行插花、压花制作等活动。

②Butterfly flowers, daisies, scarlet sages, plants in the mint family can be used to design the activity space for free picking, where children can have activities such as picking, flowers arranging, pressing and so on.

③设计一个椰糠土花池，可以让患儿自由玩土；也可给患儿提供植物移栽等活动场所。

③Design a coconut-peat flower bed to allow children to play with the soil freely; it can also be used for children to participate in activities such as plant transplanting.

9.2.5　孤独症花园案例：美国甜水圃社区花园

9.2.5　The case introduction of gardens for persons with ASD: Sweetwater Spectrum, a community garden in the United States

9.2.5.1　概况

9.2.5.1　Overview

在美国，有多达 50 万孤独症儿童正步入大龄期，这个弱势群体的总量还在不断增长，但少有住宅区适合大龄孤独症患者居住，且能满足他们的特殊需要。甜水圃是美国第一个专门为孤独症和其他相关发育障碍人群建设的社区，建筑部分由 Leddy Maytum 设计，Roche + Roche 景观公司的 Nancy 进行园林设计，社区的首批居民及其家庭成员参与建设。该项目获得 2015 年 ASLA 住宅设计荣誉奖。该社区针对孤独症的专业设计值得孤独症儿童花园借鉴。

In the United States, as many as 500 000 children with autism are entering into adulthood. The total number of this vulnerable group is still growing, but few residential areas are suitable for adult autistic patients to live and can meet their special needs. Sweetwater Spectrum is the first community built specifically for adults with autism and other related developmental disabilities in the United States. The architecture was designed by Leddy Maytum, with Nancy Roche of Roche + Roche Landscape Architecture responsible for the garden design. The first residents of the community and their family members participated in the construction. The project won the 2015 ASLA Honor Award for Residential Design. The community's professional design for autism is worthy of reference for autistic children's gardens.

甜水圃社区位于美国加利福尼亚州索诺玛，在索诺玛中心广场以西，交通便利，面积 1.13hm²。建设目标为满足孤独症患者每天特殊的生活需求，解决安全和感官问题，营造充满生机的景观环境，为居民带来有意义、有目标的生活氛围。景观设计师 Nancy 秉持的设计理念是：景观是为了丰富生活而非为了展示。

Sweetwater Spectrum is in Sonoma, California, USA, west of Sonoma Plaza, with convenient transportation and an area of 1.13 hectares. The construction goal is to meet the daily special needs of autistic patients, solve safety and sensory issues, create a vibrant landscape environment, and bring a meaningful and purposeful living atmosphere to residents. Landscape designer Nancy Roche upholds such a design philosophy: landscape is to enrich life, not to show.

9.2.5.2 布局

9.2.5.2 The layout

甜水圃大龄孤独症住宅社区为 L 形，从东往西包括社区服务中心、住宅区和有机花园三个功能区（图 9-40、图 9-41），社区花园有安静独处、娱乐活动的空间，同时还为人们提供了一个可以参与种植生产的动手实践空间。

Sweetwater Spectrum's residential community for autism adults is L-shaped. From east to west, there are three functional areas: the community service center, the residential quarters and the organic garden (Fig. 9-40, Fig. 9-41). The community garden provides space for quiet solitude and entertainment activities, while providing people with a hands-on opportunity to participate in planting and production.

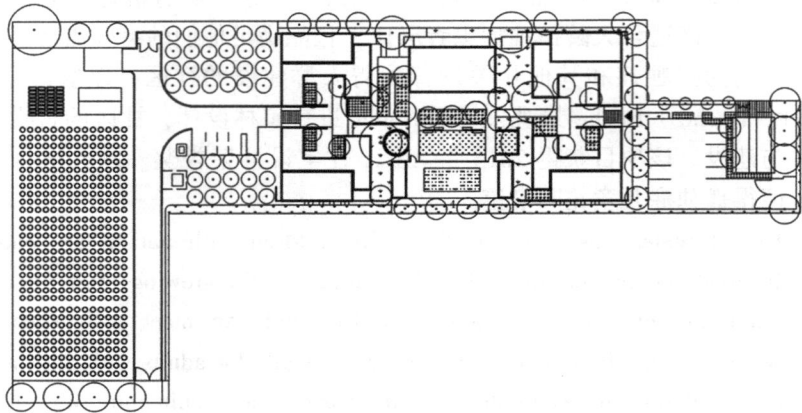

图 9-40 甜水圃平面图（李房英供图）

Fig. 9-40 Planar line drawing（by Li Fangying）

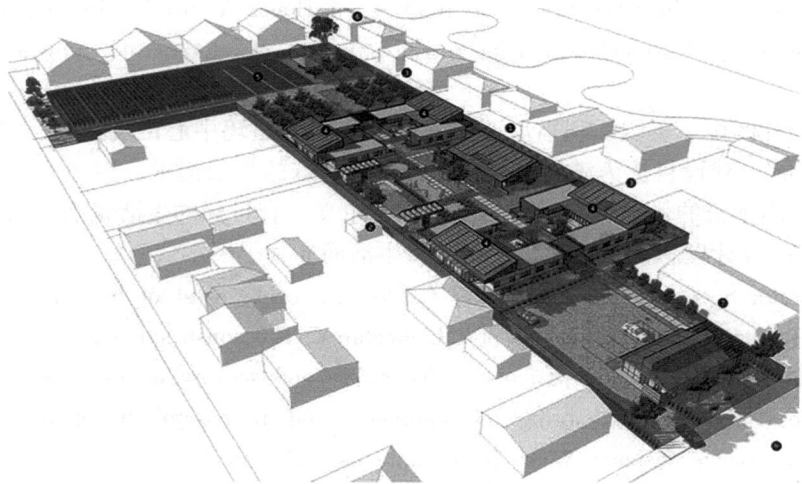

图 9-41 甜水圃孤独症住宅社区（李房英改绘）

Fig. 9-41 Sweetwater Spectrum's residential community for autism adults（by Li Fangying）

住宅区域位于社区中部，由四座环保节能居住建筑和一栋社区中心围合而成，户外空间包括共享的区域和每栋建筑独立的庭院。四栋居住建筑风格一致，卧室在两侧延伸出去形成半私密前庭空间的两翼。公共区域位于场地中心，在社区中心南面的是具有围栏的游泳池和水疗中心，中间以列植丛生禾草的草坪进行分隔，公共区域采用高1m左右的混凝土矮墙进行不同区域的划分，矮墙基部设置长凳（图9-42），成为场地的聚焦点；场地中间的榉树树冠下沿被修剪提高，使整个共享空间开敞、视线通透，便于孤独症患者和员工观察周围情况。在每个相对独立的空间里设置一些实用且有适当分隔的设施，方便居民进行轻柔的重复运动，在公共空间里居民可以进行各种的活动，如游泳、烹饪、听音乐……

图9-42　矮墙基部设置长凳（李房英改绘）
Fig. 9-42　Benches are set at the base of the wall（by Li Fangying）

The residential quarters are in the center of the community, enclosed by four environmentally friendly and energy-saving residential buildings and a community center. The outdoor space includes a shared area and a separate courtyard for each building. The four residential buildings are of the same style, with bedrooms extending out on both sides to form two wings of the semi-private vestibule. The public area is in the center of the complex; to the south of the community center is a fenced swimming pool and a spa center. The center area is divided by rows of bunch grass, while the public area is divided into different areas with low concrete walls about 1 meter high. Benches are set at the base of the wall (Fig. 9-42), which have become the focal point of the complex; the canopy of zelkova trees in the center area of the complex is trimmed high, making the entire shared space open and plainly visible, thus making it easy for autistic patients and employees to observe the surrounding situation. It is necessary to set up some practical and appropriately separated facilities in each relatively independent space to facilitate the residents in doing gentle repetitive exercises. In the public space, the residents can have various activities, such as swimming, cooking, enjoying music and so on.

在社区的西面是占地5000m²的有机农场，有果园、菜园和节能温室，以可持续方式进行栽培生产，孤独症患者可亲自参与栽培种植，收获的农产品除了自给外，还在甜水圃及索诺玛农贸市场进行对外销售。社区努力实现孤独症患者家属所期盼的生活独立和自主，达到有目标、有意义的理想生活状态。

To the west of the community is an organic farm covering an area of 5000 square meters. It has an orchard, a vegetable garden and an energy-saving greenhouse. The farm enables cultivation and production in a sustainable manner. Patients with autism can personally participate in the cultivation and planting. The harvested agricultural products can be used for self-sufficiency and sold onsite and at local farmers' market. The community strives to make patients with autism achieve life independence and autonomy expected by their families, and achieve the ideal state of living a purposeful and meaningful life.

9.2.5.3 设计细节
9.2.5.3 Design details

孤独症患者要求的环境是清晰、私密、安全和易达的。在甜水圃社区花园设计中始终在处理变化和秩序的均衡，设计师试图避免过于单调的有序。花园整体为规则式布局，中心公共空间的道路是流畅和直观的，但并非呆板的一条直线，其中加入了折线的辅助；每个居住单元的庭院布局和植物配置是相同的，但是在公共空间临近建筑的角落都有各自不同的特征，其中最受欢迎的是用树木分隔开的两个吊床的区域，非常适合人们或相聚，或独处。（图9-43）。

图 9-43 树木分隔开的吊床区域（李房英改绘）
Fig. 9-43 Hammock areas separated by trees (by Li Fangying)

Patients with autism require the environment to be clear, private, safe and accessible. The garden landscape design of Sweetwater Spectrum has been always dealing with the balance between changes and order. The designer had tried to avoid the too monotonous orderliness, and as a result, though the garden has a regular layout and the lines in the road layout of the central public space are smooth and intuitive, supplemented with broken lines, they are not rigidly straight lines; the courtyard layout and plant configuration inside each residential area are the same, but the corners facing the public space near the buildings have their own different characteristics, and the two hammock areas separated by trees are the most popular, also very suitable for people to be together or alone (Fig. 9-43).

在专门为孤独症患者设计的景观中，一定要注意孤独症患者的行为特征和使用安全，设计师 Dave Roche 在权衡利弊之后，放弃了在社区中心一侧设置砾石法式滚球场的打算；同样的，喷泉也未考虑，因为喷泉虽然可以给景观增添听觉方面的效果，但是可能会引发孤独症患者行为障碍。

In the landscape specially designed for patients with autism, attention must be paid to the behavioral characteristics of patients with autism, as well as safety in use. After weighing the pros and cons, designer Dave Roche abandoned the plan of setting up a gravel petanque court on the side of the community center; similarly, the fountain was not included in the design, because although the fountain can add auditory effects to the landscape, it can cause patients with autism to conduct disorder compulsively.

社区花园植物种植疏朗通透，有清晰柔软的纹理和适度的色彩搭配，"柔和而有个性"，植物和其他要素结合营造了具有一定变化且令人难忘的景象（图9-44）；选择耐旱、安全、无毒无害的植物。在主要树种的选择上，设计师考虑到孤独症患者对环境的可控性和简单性的特征要求，选择种植高达12m、具有明显季相变化的落叶榉树。低矮的禾草种植在宽大的种植池里，配置方式简洁，色彩对比鲜明而不复杂。庭院中只种植一种芳香植物——茉莉，以免混淆孤独症患者的嗅觉。

图 9-44　植物种植（李房英改绘）
Fig. 9-44　Combination of plants（by Li Fangying）

The plants in the community garden are well arranged with a clear view. "Soft and unique", they have clear and soft textures and appropriate color matching. The combination of plants and other elements creates an unforgettable scene with certain changes (Fig. 9-44); drought-tolerant, safe, non-toxic and harmless plants were selected, and during the selection of the main tree species, the designer took into account the characteristics of predictability and simplicity of an environment required by patients with autism, and then chose deciduous Zelkova Trees up to 12 meters high that change seasonally. The low grasses with starkly contrasting and uncomplicated colors are planted in large planting beds, arranged in a simple and clear way. Only one fragrant plant, Jasmine, is planted in the courtyard to avoid confusing the sense of smell of patients with autism.

综上所述，作为孤独症大龄患者使用的康复花园，更多考虑的是提高他们的生活品质，维持和增强自理能力，促进社会交往。甜水圃社区提供了这样的一种模式，展现出社会对孤独症患者的关怀和帮助。在其社区花园设计上充分考虑到 ASD 群体的特征和需求，在空间分隔、安全防护、路线规划、植物配置及设施安排上进行了针对性的处理。

Healing gardens for the use of adults with autism focus more on improving their quality of life, maintaining and enhancing their self-care ability, and promoting social interactions. Sweetwater Spectrum provides such a model to show the society's care and help for patients with autism. In the design of its community garden, the characteristics and needs of the ASD group have been fully considered, and targeted treatments have been carried out for space separation, safety and protection, route planning, plant configuration and facility arrangements.

思考题

根据甜水圃花园的案例谈谈孤独症康复专类园应该具备的设计要点。

Question

Please talk about the key design points of special gardens for autistic children according to the case of Sweetwater Spectrum garden.

第10章

其他类型康复花园特点及案例

Chapter 10　Characteristics and Cases of Healing Gardens for Some Groups of Patients

鉴于康复花园的针对性和功能性均较为特殊，不同人群均有各自独特的生理和心理需求。本章在前文针对主要病症患者的康复花园设计案例进行阐述的基础上，再从其他病症中选择癌症患者和肢障患者这两个相对服务频率较高的人群简要介绍其康复花园设计特点。另外，选择了新加坡城市康复花园案例和近年来康复花园中常用到的可食景观进行介绍。

Given the specific pertinence and functions of healing gardens, different populations have their own unique physical and psychological needs. After the elaboration on design cases for the patients with some major diseases in the previous chapters, this chapter focuses on the healing garden design for two groups of relatively frequent users, the cancer patients and the patients with limb disabilities. Additionally, the case of Singapore's urban healing garden and the edible landscapes commonly used in recent years are selected for introduction.

10.1　癌症患者康复花园特点及案例

10.1　Characteristics and cases of healing gardens for cancer patients

对于癌症患者而言，康复花园显得尤为重要。花园和自然对人类没有任何要求，人们可以静静地和自己的思想独处，也可以从花园的美景中寻找到宁静。所以花园可以为癌症患者提供非常舒适的治疗和康复环境，让患者在这里接受治疗，哪怕可能被告知令人震惊的诊断结果。

For cancer patients, the healing garden is particularly important because it can provide them with a very comfortable treatment and rehabilitation environment, where they receive treatment, even if they may have to face up to shocking diagnosis result. The garden and nature do not place any requirements on us, and we can be alone with our own thoughts quietly, and can also find tranquility in the beauty of the garden.

康复花园设计的一般性原则对于癌症康复花园同样适用，但是有一些特别的因素还需要单独强调。

The general principles for the design of healing gardens are also applicable to the healing gardens for cancer patients, but there are some special factors that need to be emphasized separately.

10.1.1　荫蔽
10.1.1　Shade

遮阴对所有的康复花园都很重要，但是对于癌症康复花园来说尤为重要，因为很多接受化疗的患者要求必须避免阳光直射。因此，如果一位癌症患者需要享受户外绿色空间的疗愈效果，很关键的一点就是，无论在患者需要坐下休息，还是可以步行的地方，该花园必须在一天中的任何时间都能提供荫蔽的环境。

Shade is very important for all healing gardens, but it is especially important for the healing gardens for cancer patients, because many patients undergoing chemotherapy are required to avoid direct sunlight. Therefore, if a cancer patient needs to enjoy the healing effect of outdoor green space, the crucial point is that the garden must provide a shaded environment at any time of the day where the patient needs to sit for a rest or have a walk.

10.1.2　隐私
10.1.2　Privacy

对于癌症患者，尤其是晚期患者，隐私尤为重要。花园中必须有私密或是相对隐秘的空间，供癌症患者沉思、祈祷，甚至大声哭泣。经历过化疗或者放疗的癌症患者往往会对自己的外在形象较为敏感，所以如果处于能被他人轻易瞥容的环境中，他们就会很不自在。

Privacy is particularly important for cancer patients, especially those with the prognosis of advanced cancer. There must be a private or relatively secret space in the garden for cancer patients to meditate, pray, or even cry loudly. Cancer patients who have undergone chemotherapy or radiotherapy tend to be more sensitive to their appearance, so they feel very uncomfortable in an environment where they can be easily seen.

最近研究表明，锻炼能增强癌症患者机体功能，同时能避免乏力、恶心呕吐和压抑等不良症状。因此，花园中设置散步空间，哪怕是很短的时间，也会鼓励处在治疗或者康复阶段的患者努力活动。鉴于癌症患者在化疗和放疗后容易疲乏，因此沿途需要多设置休息长椅。

Some recent studies have shown that exercise has a very positive effect on cancer patients in enhancing body functions and avoiding fatigue, nausea, vomiting, and depression. The space for walking, even for a short period of time, will encourage patients in treatment or rehabilitation to work out. For cancer patients, fatigue after chemotherapy and radiotherapy is a major symptom, so it is necessary to set up benches along the way.

除以上必须的要求以及康复花园设计的一般原则以外，对癌症康复花园而言，还有一条和其他户外疗愈空间都不同的原则，即癌症康复花园必须避免气味浓烈的花和植物。因为接受化疗的患者对气味极其敏感，任何强烈的气味，如花香、食物的气味，以及类似的特殊气味，都会引发呕吐。

In addition to the above important requirements and the general principles of the design of healing gardens, there is another principle of the healing gardens for cancer patients that is not applicable to other types of outdoor healing space. That is the healing gardens for cancer patients must avoid strong-smelling flowers and plants, because the patients undergoing chemotherapy are extremely sensitive to smell, and any strong smell, such as the smell of flowers, food, and the like, will cause vomiting.

在美国，有少数癌症康复花园会引入一些能提取抗癌药物的植物，且通常挂有植物铭牌。但目前无证据表明，癌症患者看到癌症药物提取的植物后会感到舒适。事实上，很可能恰恰相反，因为患者到癌症康复花园的目的就是想从医院的治疗中暂时逃离出来。

In the United States, a few healing gardens for cancer patients have introduced some plants, usually with plant nameplates, and those are usually plants from which anti-cancer drugs are extracted. But there is no evidence that cancer patients feel comfortable after seeing the plants from which anti-cancer drugs are extracted. In fact, it is likely that the opposite may actually be true, because the patient comes to the healing gardens to temporarily escape from the treatment in the hospital.

对于免疫系统遭到严重损害的患者，如接受骨髓移植的患者，绝对禁止暴露于土壤(比如种植床)或者是水体(如水景或者是喷泉)中。故可以设置一个能看到室内或者室外花园景观的地方作为替代。

For the patients with severely compromised immune systems, such as the patients receiving bone marrow transplants, exposure to soil (such as planting beds) or water bodies (such as waterscapes or fountains) must be completely forbidden. In the above situations, it is very necessary to set up a location with a view of the indoor or outdoor garden as an alternative.

10.1.3　案例：癌症生命线屋顶花园
10.1.3　Case studies：the roof gardens of Cancer Lifeline

10.1.3.1　设施简介
10.1.3.1　Introduction of facilities

美国西雅图多萝西·奥布莱恩中心癌症生命线是一家私人的非营利、非医疗机构，它面向所有经历癌症治疗的患者。该中心于 1999 年开业，由私人捐助成立，所有的服务均免费。中心位于西雅图居民区，为一处安静的两层小楼，是由杂货铺和公寓重新设计和改造而成。人们到这里来寻求情感上的支持，参与团队活动和课程，使用图书馆以及享受在楼顶的三个屋顶小花园。

Cancer Lifeline at Dorothy S. O'brien Center in Seattle is a private, non-profit, non-medical organization that caters to all patients experiencing cancer. The center opened in 1999 and was established through private donations. All services there are free. It is in a quiet two-story building in a residential area in Seattle. The building used to be a grocery store and apartment, but has now been redesigned and remodeled. People come here to seek emotional support, participate in team activities and courses, use the library, and enjoy the three small roof gardens on the top of the building.

10.1.3.2　设计理念
10.1.3.2　The design philosophy

这座花园由华盛顿大学一群大四的景观设计本科生，在为期十周的课程中设计和建造完成，其指导教师是丹尼尔·温特伯顿教授。师生们和中心的员工以及使用者们紧密协作，很好地利用屋顶空间打造出一处简单、富有教育意义且经济实惠的花园，以满足使用者的需求。花园位于室外，但同时又受到保护，拥有冥想空间，有嗅觉体验以及游玩功能。

The garden was designed and built by a group of senior students majoring in landscape architecture from the University of Washington in a ten-week course. The instructor was professor Daniel Winterbottom. The teacher and students worked closely with the staff and users of the center to create a simple, instructional and affordable garden to meet the needs of users while making good use of the roof space. The design requires that the garden be located outdoors while being shielded, so as to provide space for meditation, elements of smell, and the function to enable strolling about.

10.1.3.3　室外空间
10.1.3.3　Outdoor space

该花园分为三个两层的屋顶小花园，其中两个邻近建筑边缘，一个稍微靠近中心。

The garden is divided into three two-story roof gardens, two of which are adjacent to the edge of the building and one of which is slightly close to the center.

(1)冥想花园

（1）The meditation garden

冥想花园面积约为 13m²，全部采用木质地板，整个花园与建筑物的两面围墙贴合起来，另外一面可以通过竹篱看到旁边的建筑。花园中有一个小型旋转喷泉，设置在鹅卵石中，对面是一张直角坐凳和两把可移动园椅。旁边还设有一个小的陶瓷壶，上面的标识写道："这是你自己的花园。请将开败的花朵，或是掉落的花瓣，或者是任何您看到的杂物放入壶内。花园由衷地感谢您。"在花园里可以欣赏喷泉，听潺潺水声，感受原木地板松软的质感，还有摇曳的竹篱，多样的绿色（但没有真正的花），以及一些园艺摆件（鸟巢、马赛克等），这一切使整个花园拥有一种非常放松的感觉，可以让人在此静思。花园上部还有一个半遮蔽的玻璃屋顶，可以在天冷的时候提供保暖，以保证人们的舒适感。

The meditation garden covers an area of about 13m², all of which adopts wooden floors. The entire garden and the two walls of the building form closed space, and the adjacent building can be seen through the bamboo fence from another side. There is a small revolving fountain, set in pebbles in the garden; the opposite is a right-angle stool and two movable garden chairs. There is also a small ceramic pot next to it, and the sign on the pot says: "This is your own garden. Please put the withered flowers, fallen petals, or any debris you see into the pot. The garden sincerely appreciate it." In the garden, people can enjoy the fountain, hear the gurgling water and feel the soft texture of the wood floor; there are also swaying bamboo fences, diversified greenery (but no real flowers), and some gardening decorations (the bird's nest, mosaics, and so on), all of which make the whole garden relaxing, allowing people to quietly meditate there. There is also a glass roof above that partially shields the garden, providing warmth in cold weather to ensure people's comfort.

(2)庆祝花园

（2）The celebration garden

庆祝花园位于角落边，面积约为 310m²，其两边是二层的房间，另外两边透过栅栏可看到周边邻居的房屋。花园中设置木质棚架，下面有一些固定座位和多种样式的可移动园椅，以及带遮阳伞的桌子。这部分空间有户外咖啡吧的感觉，是专为取得一定康复效果后的患者社交活动而设计的。花园中种植薰衣草、迷迭香、牛至、鼠尾草、百里香和薄荷等。只有在人非常靠近或者按压叶片时才会闻到气味。

The celebration garden is located on the corner with an area of about 310m². There are two-story rooms on its two sides, and the houses of surrounding neighbors can be seen through the fences on the other two sides. There are arbour-style wooden flower stands in the garden, underneath which are some fixed seats, movable garden chairs of various styles, and tables with parasols. Especially designed for social life following some progress in recovery, this place has the feel of outdoor cafe. Lavender, Rosemary, Oregano, Sage, Thyme and

Mint are grown in the garden. the plants selected in the garden only smell when people are very close to them or when their leaves are being pressed.

(3) 天地园

(3) The earth and sky garden

天地园从二层主厅的走廊可以直接看到，入口处是一扇风格独特的树形铁门，从庆祝花园和多功能艺术厅都可以到达这里。花园里有花台边缘长凳、可移动园椅、贵妃椅等可供休息。顶上还有一个特色木质架子，整个花园空间呈"L"形，树形意象的铁门可以半开，从而将空间分隔成两部分。园内栽植的玉簪、钓钟柳、黄精、天竺葵、矮松等植物则为花园提供了多样的色彩和质感。一侧栽植的植物低矮，以便看到旁边的走廊；另一侧的植物则高一些，以保证私密性。遮阴篷布则在夏天提供了一定程度的荫蔽。

The earth and sky garden can be seen directly from the corridor of the main hall on the second floor. There is a tree-shaped iron door, quite a unique style, at the entrance, which can be reached from the celebration garden and the multifunctional/art hall. In the garden, there are flower bed curb-turned benches, movable garden chairs, chaises longues, and so on. With a characteristic wooden shelf above it, the space in the entire garden is L-shaped, and the tree-shaped iron door can be half-opened to divide the space into two parts. The fragrant plantain Lilies, Beardtongues, Solomon's seals, Geraniums, Virginia Pines and other plants planted in the garden provide a variety of colors and textures for the garden. The plants on one side are low so that people can see the corridor next to them, and the plants on the other side are taller to ensure privacy. The shade tarps provide some shade in summer.

10.1.3.4　花园使用说明

10.1.3.4　Garden instructions

冥想花园一般是供一人独处或两人安静的交谈和沉思的场所。有时一个人也可以独自在花园中静坐冥想。因为花园很私密，空间也不大，基本只能容纳一人。

The meditation garden is generally for a person to be alone or for two people to talk and meditate quietly. Sometimes a person can practice qigong alone in the garden. The garden is very private and its own space is not large, so if one person is in the garden, another person is unlikely to enter it.

庆祝花园在中午会有阳光，因此通常供工作人员或其他人午餐或者小型会议使用。在最初的设计中，有一个供攀缘植物攀爬的铜管凉亭，但并未实施，因此目前花园中的光线相对较好。就西雅图多雨的气候来说，这反而成为一个优点。

The celebration garden has sunshine at noon, so it is usually used by staff or others for lunch or small meetings. In the original design, there was a copper-tube pavilion for climbing plants to climb, but it was not implemented afterwards, and as a result, the light in the garden is better than expected. Given the rainy climate in Seattle, this is an unexpected advantage.

天地园的目的是让人感觉到有所把控。因为患者通常会感觉他们的身体和生活已经不受控制。这里可以用作气功练习和感知自然的场地。因为紧靠着艺术室，也可以进行艺术创作和创意写作课程。从花园中采摘的植物和花朵可以供插花课程使用。

The purpose of the earth and sky garden is to make people feel they are in control, as the patients usually feel that their body and life have been out of control. The garden can be used to as a venue for practicing qigong and sensing nature. It is close to the art room, so art and creative writing courses can also be provided. Plants and flowers picked from the garden can be used for flower arrangement courses.

10.1.3.5　花园的益处
10.1.3.5　Main advantages

①植物、材料和家具都很低调，规模适宜，也适合当地环境。

①Plants, materials and furniture are very low-key, suitable in size and for the native place.

②将屋顶空间分隔成三个独立的花园，可以使不同的人群或个人都能找到属于自己的私密空间。

②The roof space is divided into three independent gardens, so that different groups of people or individuals can find their own private space.

③营造了不同感受的活动空间和多元化氛围。

③Different feels of space provide diversified atmosphere for activities.

④花园为二层的内部空间提供了绿色的景观(廊架和多功能屋)。

④The garden provides a green landscape (the gallery frames and multi-function house) for the internal space on the second floor.

⑤提供获得主动控制感受的机会(如选择三个不同空间、挪动桌椅、捡拾落花等)。

⑤For garden users, it provides a variety of options to quietly gain a sense of control (choices in three different places, movable garden chairs, picking up flowers and plant residues), and to experience private space.

10.1.3.6　花园不足之处
10.1.3.6　Main disadvantages

①庆祝花园过于临近街道、邻居。

①The celebration garden is a little too open to the streets, neighbors, and traffic.

②雨天木质地板很滑，存在一定安全隐患。

②Wooden floors become slippery on rainy days, causing certain hidden dangers.

10.2　肢障者康复花园设计要点简介

10.2　The introduction to healing gardens for persons with limb disabilities

肢障患者由于肢体残损而导致身心功能产生严重障碍，不但个人生活不能自理，而且影响社会交往和工作。本节针对上肢残障和下肢残障两类患者的共性和差异对其康复花园设计要点进行简要介绍。

Persons with disabilities, due to disabilities or severe disabilities and severe physical and mental dysfunctions, cannot take care of themselves in personal life. What's worse, their participation in social life and work is also affected. This section briefly introduces the key design points of the healing garden for patients with upper and lower limb disabilities according to their commonalities and differences.

10.2.1　肢障患者的特点

10.2.1　Characteristics of persons with limb disabilities

针对肢障人员的康复花园，应充分考虑到其肢体能力问题。在进行园地设计时，和其他康复花园一样，除应同时考虑其肢体锻炼和情绪康复的目标之外，还应考虑上肢残障人士由于其操作难度较大，不适合精度要求高的手工类活动；下肢残障人员则应充分考虑其下部肢体移动能力不足。设计难度不大的园艺操作，让他们的疗愈目标聚焦情绪康复和简单肢体锻炼。

Persons with limb disabilities should fully consider the problem in their limbs abilities. When designing the garden, designers should consider emotional rehabilitation and extremity exercise goals of those patients. Operation is difficult for people with upper limb disabilities and they are not very suitable for hands-on activities. For people with lower limb disabilities, their limited mobility in their lower limbs should be fully considered, but the horticultural operation is not too difficult. So their healing goals are emotional rehabilitation and simple extremity exercises.

10.2.2　肢障患者专类园设计要点

10.2.2　Key design points of specialty gardens for persons with limb disabilities

（1）地形

（1）Terrains

应避免台阶通道，所有场地应具有比较开阔的视野，行进路线无障碍，且适合设计环形类汇聚交流场所；地面应防滑，尤其要考虑在雨雪天气情况下的地面状况。

Passageways with stairs should be avoided, and all venues should provide a relatively wide field of vision, so that the road ahead can be seen; the ground should be anti-skidding, and the ground conditions in rain or snow should be especially considered; it is reasonable to design ringlike venues for gathering and exchanges.

（2）水面

（2）The water surface

水面可以使人安静，通常设计为浅水池，旁设座椅，供肢障者休息。

The water surface can calm people, and shallow water can be designed with seats next to it for rest.

（3）环境设施

（3）The environment and facilities

在开阔性场地设计小型交流环境，以便于锻炼的人员相互沟通。在观赏区设置遮雨棚并设计合理的休息位置，应在锐角、直角物体外部进行防撞处理，以防止事故发生。针对上肢残障人士，可以设计便于下部肢体锻炼的设施，如脚底按摩步道类；针对下肢残障人士要考虑在座椅处设置扶手，便于其起立；在主活动区设置园艺操作间，操作台应具有代步车停靠和锁闭装置，并符合操作人员坐代步车行动的高度条件。

Set up a small environment for communication in the open field to make it easy for the people who exercise to communicate with each other. Set up some shelters in the viewing area to provide a quiet healing environment. Positions for rest should be reasonably designed, and anti-collision treatments should be carried out on the exterior of acute-and right-angle objects to prevent accidents. And for people with upper limb disabilities, design some equipments that is convenient for lower limbs exercise, such as reflexology foot paths. At the same time, for people with lower limb disabilities, armrests should be provided at the seats to facilitate them in standing up; a gardening operation room should be set up in the main activity area, and the operating desk should be equipped with devices that enables the parking and locking of scooters and should meet the height requirements so that the patients can operate when sitting on a scooter.

（4）植物

（4）Plants

可选择一些长期开花的植物，或者蜜源植物，配合招鸟工程，建设香草瓜果类专项观赏园，或豢养一些鸟类，使人们可以在花园内聆听鸟鸣。也可提供部分可移植的苗木，组织团队型园艺活动，提升成就感。针对下肢残障者可设计移栽区和采摘区，种植如茉莉、月季、叶子花等适合修剪的植物，以便于开展修剪或插花活动。

Some plants with long-lasting flowers can be selected, and special ornamental gardens such as herb gardens can be built. Designers can design some bird-attracting projects or raise some birds, so that people can enjoy the birdsong in the garden. They can also supply some seedlings for transplantation and organize team gardening activities to promote people's sense of accomplishment. For people with lower limb disabilities, transplanting and picking areas can be designed; cultivate plants such as jasmine, rose, bougainvillea and other plants easy to prune for pruning or flower arrangement activities.

10.3　社区共建城市绿地（CIB 项目）

10.3　Community co-construction of urban green（CIB programme）

　　鉴于花园城市新加坡在康复花园的设计建造领域有诸多先行的经验，因此本节着重介绍其做得较好的社区共建城市绿地项目。新加坡自建国以来就非常重视城市绿地建设，沿着道路、公共用地和住宅区均密集种植树木，并创建了许多公园和绿地。过去几十年里，其绿色基础设施逐年增加和完善，尤其是社区共建的生物多样性绿地，让新加坡变成大自然中的城市（新加坡国家公园委员会网站）。为了鼓励和促进社区中的园艺文化活动，新加坡于 2005 年启动创建了 CIB 社区绽放计划。该项目旨在将园艺推广为一种生活方式，并最终发展成为全国性的园艺运动。这些 CIB 社区花园主要位于公共和私人住宅区域，但越来越多的教育机构、工业和商业区域也开始建造自己的花园，并取得了非凡的成果，包括：邻里之间有了更大的社会凝聚力，师生有更多机会参与有意义的户外活动，退休人员通过园艺活动保持活力。

　　In view of the fact that Garden City Singapore has many pioneering experiences in the field of design and construction of rehabilitation gardens, this section focuses on its better community co-construction of urban green space gardens. Since the founding of Singapore, it has attached great importance to the urban green space. Trees were planted intensively along roads, public and housing estates, as well as the creation of many parks and green spaces. Over the decades, green infrastructures have been increased and enhanced greater emphasis was placed on enhancing the biodiversity of green spaces and engaging the community in sustaining these green efforts, hence transforming Singapore into a City in Nature（The website of Singapore National Parks Board）. To encourage and promote a gardening culture in the community, the Community in Bloom（CIB）programme was created. Launched in 2005, the programme aims to promote gardening as a lifestyle choice and eventually becoming a national gardening movement. These CIB community gardens are located largely in public and private residential estate, but increasingly, educational institutions, industrial and commercial estates have also begun to start heir own gardens. Since its inception, many positive outcomes have been shown, including greater social cohesion between neighbours, teachers and students engaging in meaningful outdoor activities, and also retirees staying active through gardening

　　因为生机盎然的自然环境是患者康复过程的一个重要促进因素，所以新加坡医疗保健机构也逐渐将活跃的花园空间纳入医疗保健环境，这一举措为患者和医院工作人员带来了诸多福利。其中，首批围绕亲生物设计原则进行规划和设计之一的邱德拔医院，无疑是医院康复花园早期的雏形。

As a vibrant natural environment is an important contributor to the healing process of patients, Singapore healthcare institutions are increasingly incorporating active garden spaces into their healthcare environments, which brings many benefits to patients and hospital staff. Among them, Khoo Teck Puat Hospital, which is planned and designed around the principle of biophilic design, is undoubtedly the early prototype of the hospital rehabilitation garden.

2010 年建成开业的邱德拔医院是一所公立医院，被誉为"花园中的医院和医院中的花园"。彼时治疗性花园设计的原则尚未被明确规定在绿地设计中，但邱德拔医院的发展却引领人们意识到人与植物互动的益处。医院场地的中心是一个森林般的花园，里面有一个池塘作为雨水集水区，周围树木郁郁葱葱，让患者、工作人员和游客等都能享受到大自然的康复益处（图 10-1）。花园包含了视觉、嗅觉、听觉、动物、植物多样性五大原则，并在蓝绿空间内整合了社区空间。除了为医院提供风景和人们的娱乐空间，花园也行使了生态循环的功能。下雨时，池塘收集医院的地面地表径流，然后通过水泵返回水灌溉医院植物（图 10-2）。该院落的设计还包括八个具有不同种植主题和用户入口的屋顶花园，其中部分对公众和种植水果蔬菜的志愿者开放；而另一部分则为管控式出入，仅供医院员工和患者使用，因为要保障患者（尤其是体弱的老年人和失智症患者）的安全。医院开业十几年来，屋顶花园不仅为厨房提供蔬菜，多余的还能售卖营利。

Opened in 2010, Khoo Teck Puat Hospital is a public hospital known as "a hospital in a garden, and a garden in a hospital". At the time the principles of therapeutic garden

图 10-1　从邱德拔医院俯视新加坡义顺池公园鸟瞰图（图片来源：维基共享资源）

Fig. 10-1　Aerial view of Khoo Teck Puat Hospital overlooking Yishun Pond Park, Singapore（Photo credit：Wikipedia Commons）

图 10-2　新加坡邱德拔医院中心水景（图片来源：Pong Junxiang）

Fig. 10-2　View of central water feature of Khoo Tech Puat Hospital, Singapore（Photo credit：Pong Junxiang）

design were still not specifically included in the design of the green spaces, but the development of Khoo Teck Puat Hospital led to an awareness of the benefits of human-plant interaction. At the heart of the hospital complex is a forest-like garden that showcases a pond amidst lush and verdant surrounding, giving both patients, staff and visitors alike access to the restorative benefits of nature (Fig. 10-1). Gardens encompass the five principles of sight, smell, sound, diversity of flora and fauna, and incorporating community spaces within the blue-green areas. Besides providing a scenic view for the hospital as well as recreation space for people, when it rains, the pond collects surface runoff from the hospital grounds, then returns the water back through pumps to irrigate the plants in the hospital (Fig. 10-2). The design of the compound also includes 8 roof gardens with different planting themes and user access, some open to the public and volunteers who tend to the planting of fruits and vegetables, while others have controlled access where only hospital staff and patients can enter to ensure the patients' safety, particularly the frail elderly and the patients with dementia. Over a decade since the hospital's opening, the roof gardens are thriving. Harvest from the vegetable plots used in the kitchen and excess are sold for profit.

　　邱德拔医院成功开创了以亲生物方式设计医疗保健环境的先例，其强调为患者创造一个安全有益的康复环境，为后来更多针对特定使用者的康复花园设计建造奠定了基础。

　　The success of Khoo Teck Puat Hospital set a precedent for designing healthcare settings with a biophilic approach, and the emphasis on creating a safe and conducive environment for patients to enjoy and recover. This in turn, paved the way for more of such user-specific green spaces such as therapeutic gardens.

10.3.1　为老龄化社区建立园艺康健花园
10.3.1　Building a therapeutic horticulture approach for the ageing community

图 10-3　新加坡的康健花园（黄印星　摄）
Fig. 10-3　Therapeutic horticulture garden at HortPark, Singapore(by Huang Yinxing)

　　基于 CIB 社区花园项目的成功，以及在医疗保健环境中强调绿色植物的积极成果，新加坡计划创建一个示范花园，以促进老年人和失智症患者进行治疗性园艺活动。2016 年，该花园在位于新加坡西南部的霍特公园和一站式园艺生活方式中心推出（图 10-3）。花园的发展汇集了来自国家公园委员会的设计师和研究人员和国立大学医院的老年病学专家，以及老年和失智症护理领域的各种专家。与此同时，该团队还参与了一项关于园艺康健对亚洲老年人心理健康影响的研究。在这项研究中，69

名老年参与者被随机分成小组实施园艺康健，或者在候补名单中作为对照组。其中的一些活动包括室内和室外园艺活动，以及参观公园，如新加坡植物园、裕华园和滨海湾花园的云雾森林和花穹。这些疗程每周进行 1 次，持续 12 周，随后每月进行 1 次，持续 3 个月。在整个疗程中，通过参与者对抑郁和焦虑症状的自我报告、社会联系度和心理健康度以及免疫标志物测试，对参与者的心理健康进行评估。这项研究在很大程度上取得了积极的成果，园艺康健组的参与者表现出更高的生活满意度和社会联系度。这也反映在视觉空间技能、记忆力和语言流畅性等方面(Ng，2018)。

Building on the success of the CIB Community Gardens movement and the positive outcomes of emphasizing greenery in healthcare settings, a plan was drawn to create a prototype garden to promote therapeutic horticultural activities for the elderly and those with dementia. In 2016, the garden was launched at HortPark, a public park and one-stop gardening lifestyle hub located in the southwestern part of Singapore (Fig. 10-3). The development of the garden brought together the designers and researchers from the National Parks Board, gerontologist from the National University Hospital, and various experts in the field of elderly and dementia care. In conjunction with this development, the same team was also involved in a study on the effects of horticultural therapy on Asian elderly's mental health. In the study, 69 elderly participants were randomly assigned to receive horticultural therapy in groups or to be in the waitlist, which served as a control group. Some of the activities include indoor and outdoor horticultural activities and visits to parks, such as the Singapore Botanic Gardens, Chinese Garden, the Cloud Forest and Flower Dome at the Gardens by the Bay. The sessions were conducted weekly for 12 weeks, and subsequently every month for three months. Throughout the sessions, the participants' mental health was assessed through self-reports of depressive and anxiety symptomatology, social connectedness and psychological well-being and tests on immunological markers. The results of this study are largely positive, with participants in the horticultural therapy group showing greater life satisfaction and social connectedness. This is also reflected in areas such as visuospatial skills, memory and verbal fluency (Ng, 2018).

10.3.2　公园内的康复花园
10.3.2　Healing gardens at public parks

在霍特公园的示范花园取得成功之后，新加坡国家公园局开始开发覆盖全国公园的康复花园网络。前两个康复花园是在 2017—2018 年，分别位于中峇鲁公园和碧山—宏茂桥公园内。这些公园被选作新花园的原因，是它们位于拥有成熟住宅区的城镇中，老年居民的比例很高。同时由于老年居民的比例较高，附近也有更多的老年护理设施和老年活动中心。因此，将康复花园设在这些设施附近，可以让更多的人享受和受益。在设计阶段，开发团队访问了附近的养老院和老年护理中心，以收集目标用户的需求信息。此外，开发团队还向老年学专家、心理学家和从事老年人和失智症护理工作的职业治疗师寻求专家建议。

Following the success of the prototype at HortPark, the National Parks Board began to develop a network of healing gardens throughout publics parks in Singapore. The first two healing gardens were developed in Tiong Bahru Park and Bishan-Ang Mo Kio Park respectively between 2017 to 2018. These parks were chosen as sites for the new gardens as they are located in towns with many mature housing estates where the percentage of elderly residents are significant. Due to the high percentage of elderly residents, more elder care facilities and senior activity centres can also be found around the vicinity. Hence, locating the healing gardens near these facilities allows for more people to enjoy and benefit from them. During the design stage, the development team visited nursing homes and elder care centres within the vicinity to gather information on the needs and requirements of target users. Expert advices were also sought from gerontologist, psychologists and occupational therapists working on elderly and dementia care.

考虑到以下几个因素，建造康复花园的场地一般大小为 800~1000m²。首先，考虑到老年人使用时的身体限制，花园选址应根据老年人与车辆下客区和附近设施(如厕所、洗手区和自动饮水器)的邻近程度来决定。其次，为了便于移动，特别是为了方便坐轮椅人士的移动，选择相对平坦的地形，从而在花园内尽可能最大化实现各种"人—植"互动和体验。再次，新加坡是热带气候，昼温可达 30~33℃，利用现有成熟树木提供充足的荫蔽至关重要。研究表明，与无遮阴的室外相比，室外有树木遮阴区域的温度可显著降低 2.5℃。最后，这两个康复花园均位于现有公园内，可借用绿色植物来增强花园景观体验。此外，碧山—宏茂桥公园还毗邻一个湖泊，为使用者提供了欣赏风景的有利位置(图 10-4、图 10-5)。

图 10-4 从新加坡碧山—宏茂桥公园的康复花园入口所看到的景色(谭凯新 摄)
Fig. 10-4 View from entrance to healing garden at Bishan-Ang Mo Kio Park, Singapore (by Tan Kaixin)

The sites for the healing gardens, ranging from the size of 800-1000m², were selected according to several factors. Firstly, considering the physical limitation of elderly users, sites were selected based on their proximity to the vehicular drop-off areas and nearby amenities such as toilets, washing areas and drinking fountains. Secondly, relatively flat terrains were chosen for the ease of movement, especially for people on wheelchair, thus maximizing a variety of people-plant interactions and experiences within the garden enclosures. Thirdly, in the tropical climate of Singapore where the day temperature can range between 30-33℃, adequate shade in the form of existing mature trees is crucial to provide respite for people

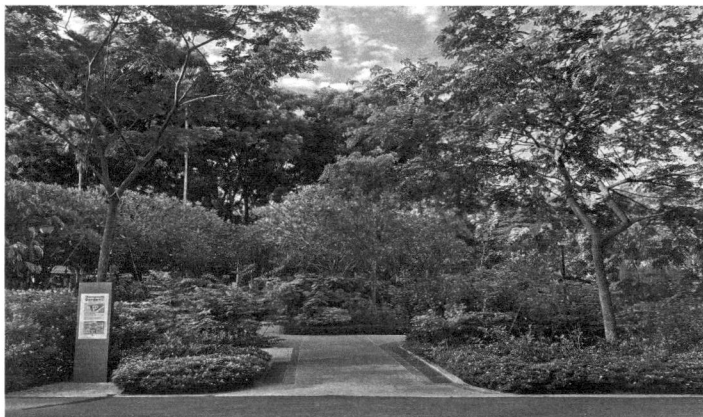

图 10-5　新加坡中峇鲁公园的康复花园入口景色（谭凯新　摄）
Fig. 10-5　Entrance view of healing garden at Tiong Bahru Park, Singapore(by Tan Kaixin)

from the sun. Studies have shown that the temperature in outdoor area shaded by trees can be significantly reduced by 2.5℃ as compared to that in an unshaded outdoor area. Lastly, both sites are already located within existing parks, so borrowed greenery become advantages to enhance the landscape experience in the gardens. Furthermore, the site at Bishan-Ang Mo Kio Park is also located adjacent to a lake, thus providing a vantage point for users to admire the scenic view (Fig. 10-4, Fig. 10-5).

在为老年人和失智症患者设计循环路径时，道路设计为简单而清晰的布局。这两个花园的循环路径都由闭环道路组成，循环路径采用"8"字形，允许用户轻易地穿梭于各个空间。在为失智症患者设计时，特别为阿兹海默病患者设计时，这种路径设计至关重要，因为患者往往丧失了认知能力，导致他们四处游荡、迷路，对自己的位置感到困惑。有了闭环路径，失智症患者就能够安全地在花园周围移动；同样地，也让陪伴他们的护理人员能够知晓他们在花园的行踪。除此之外，花园的边界轮廓分明，用屏风式植物创建边界，其他不同形式和材质的植物则用以软化边缘。这为用户提供了安全和隐私。这些花园内部还建造了被动区和主动区，使人们的注意力远离花园空间外的其他干扰。为了鼓励人们与植物亲密接触，沿途种植了不同色彩、纹理和趣味特征的植物。在某些区域，还引进了高床花槽。例如，碧山—宏茂桥公园的康复花园以一个倾斜的高床花槽为特色，供不同能力的使用者与不同高度的植物互动(图 10-6、图 10-7)。另外，高床花槽的设计也方便轮椅使用者参与简单的园艺活动，如播种、修剪和浇水，且每个花园中的高床花槽旁边都有一个洗手区。

In designing the circulation with elderly and those with dementia in mind, pathways are set with simple and clear layouts. The circulation in both gardens consist of closed loop paths, in the form of a figure-of-eight to allow users to navigate easily through the spaces. This is a crucial design for people with dementia, especially those with AD where they tend to lose their ability to recognise familiar places, resulting in them wandering around, getting lost and confused about their location. By having a closed loop path, people with dementia will be able to move around the garden safely and likewise, the caregivers accompanying them will be

图 10-6 新加坡碧山—宏茂桥公园的康复花园中的倾斜种植槽（谭凯新 摄）

Fig. 10-6 View of sloped planters at healing garden at Bishan-Ang Mo Kio Park, Singapore（by Tan Kaixin）

图 10-7 新加坡碧山—宏茂桥公园康复花园内易于轮椅使用者操作的种植槽（谭凯新 摄）

Fig. 10-7 View of wheelchair accessible planters at the healing garden at Bishan-Ang Mo Kio Park, Singapore（by Tan Kaixin）

assured of their whereabouts within the garden. In addition to this, the perimeters of the gardens are well defined, with screening plants to create boundaries and other plants of various forms, and textures to soften the edges. This provides both security and privacy for the users. Within the gardens, passive and active zones are also created to redirect the attention away from other distractions outside of the garden spaces. To encourage people to get into close contact with plants, plants of different colours, textures and interesting features such as seed pods are planted along the paths. In certain areas, raised planters were also introduced. For example, the healing garden at Bishan-Ang Mo Kio Park features a sloped raised planter for users of different abilities to interact with plants at different heights (Fig. 10-6, Fig. 10-7). The first type is designed to allow for wheelchair bound users to be involved in simple horticultural activities like sowing seeds, pruning and watering of plants. A washing point next to the raised planters is provided at each of the gardens.

康复花园的另一重要功能是举办活动。因此，花园中包含了多个可容纳 10~12 人的大型庇护间。除了让使用者进行休息和社交，庇护间还可举行团体活动，包括康健园艺会议等。此外，花园配备了健身器材和轮椅，以鼓励体育锻炼。结合这些有意义的活动可提高人们使用花园的意识和频率(图 10-8、图 10-9)。

One important aspect of healing gardens is the opportunity for scheduled and programmed activities to take place. Hence, large shelters that can accommodate 10 – 12 people are included into the gardens. Besides allowing users to rest and socialize, the shelters also allow group activities, including therapeutic horticulture sessions to take place. Additionally, fitness equipment, some wheelchairs are also incorporated into the garden to encourage physical exercise. This increases the awareness and frequency of the garden use with meaningful activities (Fig. 10-8, Fig. 10-9).

图 10-8　新加坡中峇鲁公园中康复花园种植运动区(谭凯新　摄)
Fig. 10-8　View of plantings and exercise area at healing garden at Tiong Bahru Park, Singapore (by Tan Kaixin)

图 10-9　新加坡中峇鲁公园康复花园内带栏杆的小路(谭凯新　摄)
Fig. 10-9　View of pathway with railings at healing garden at Tiong Bahru Park, Singapore (by Tan Kaixin)

精心设计的花园空间满足了老年人和失智症患者的需要，还为公众提供了娱乐用地。中峇鲁公园和碧山—宏茂桥公园中的康复花园，成了在其他公园和医疗保健设施(如养老院)中建造更多康复花园的典范。将康复花园纳入这些公园和医疗保健设施对优化治疗环境具有积极意义。

With carefully designed spaces in the gardens catering to the needs of the elderly and those with dementia, as well as providing another destination for the general public to enjoy, the healing gardens at Tiong Bahru Park and Bishan-Ang Mo Kio Park set an important precedent for more of such gardens in other parks, as well as in healthcare facilities, such as nursing homes, where the incorporation of healing gardens within its compounds will greatly enhance the healing environment.

10.4　可食地景设计
10.4　The design of edible landscapes

较之大多数由观赏植物构成的康复花园，可食地景是近年来应用于康复花园的一种特殊景观表现形式。可食地景是指有生产性的、由可提供食用的果树蔬菜等植物构成的景观。近些年，随着都市人群的返璞归真和对田园生活的向往，可食地景的出现为人们在紧张的城市生活中找到了一种城市景观建设与农业发展相兼容的新型景观发展模式。可食用地景可利用有限的、零散的土地挖掘最大化的空间使用潜力，满足人们农耕、食用、观赏休闲的景观体验，并在拉动城市农业经济发展的同时，照护城市居民健康。可食用植物的生长过程，令城市居民感到无比新奇和兴奋，他们在这里经历播种、收获，可以尝试种植不同的植物，根据自己的需要选定种植数量与种植时间，这些区域的植物规模、色泽、质感，甚至味道都会发生季节性变化（图 10-10 至图 10-14）。可食地景概念的提出，是一次重新思考人们应当种植什么的观念意识复兴，可谓城市人群与植物互作关系的螺旋式上升。

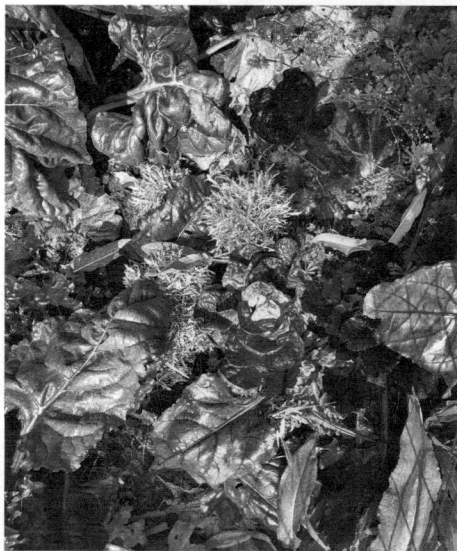

图 10-10　晨曦中的各种紫色（袁惠燕　摄）　图 10-11　紫色和绿色羽衣甘蓝（袁惠燕　摄）
Fig. 10-10　Various purples in morning light　　Fig. 10-11　Purple and green Kale
（by Yuan Huiyan）　　　　　　　　　　（by Yuan Huiyan）

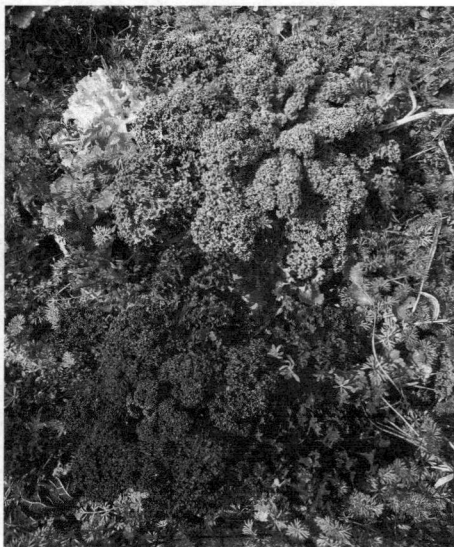

Compared with most rehabilitation gardens composed of ornamental plants, edible landscape is a special landscape expression used in rehabilitation gardens in recent years, which is a productive landscape consisting of fruit trees, vegetables and other edible plants. In recent years, with the urban population's return to nature and yearning for rural life, the emergence of edible landscapes has found a new landscape development model that makes the urban landscape construction compatible with agricultural development in the busy urban life. The edible landscape can use the limited and scattered land to tap greater potential for space use, provide people with landscape experiences in farming, eating, and viewing and entertainment, and bring health care to urban residents while promoting the development of the urban agricultural economy. The growth of edible plants extremely amazes and excites the city dwellers. They experience sowing and harvesting here, and they can try planting different plants, grow more plants or less according to their needs, and arrange the planting time as they like. The size, color and texture and even taste of the plants in these areas change seasonally (Fig. 10-10 to Fig. 10-14). The concept of the edible landscape is a revival of the mentality of rethinking what we should grow and can be

图 10-12　高低错落的绿（袁惠燕　摄）
Fig. 10-12　Greens in different forms
（by Yuan Huiyan）

图 10-13　早春多彩的花园（袁惠燕　摄）
Fig. 10-13　Colorful garden in early spring
（by Yuan Huiyan）

described as a spiral upgrade in the interaction between the urban population and plants.

　　与其说可食地景是新事物，不如说它是传统农耕文化的复兴。史上已有将农作物和装饰性植物共同种植的案例。在古巴比伦和古埃及的园林中都使用过可食地景。早在 10 世纪，意大利本笃会的僧侣就在月季园的外围种植草药。直到文艺复兴时期，人们才开始有意识地将农作物和纯粹装饰性的植物分开种植。可食地景很强调创意，精心设计的菜园可以很惊艳。配置果树、蔬菜、药用植物，特别是可食的开花植物，可食地景能让城市变得更加美丽和美味。改造城市地景，在城市居民区种地，对于一些西方国家而言早已不是新鲜事。从社区菜园到学校农园，都能让城市人享受晴耕雨

图 10-14　花期中的甘蓝也很美（袁惠燕　摄）

Fig. 10-14　Cabbage in flower season is beautiful（by Yuan Huiyan）

读的惬意。而且，随着朴门永续农业设计理念的日益普及，将自家本来平淡无趣的草坪改造成美观又美味的可食地景更是趋势所在。20 世纪 80 年代，园林设计师、环保主义者罗伯特·科瑞克提出了有趣的术语"可食地景"，它代表着园林设计与农业生产的融合。随后，罗丝琳德·克瑞丝在 1982 年出版了《可食地景完全指南》一书，将这一概念引入主流园林设计界。可食地景不是简单的种地，而是用设计生态园林的方式设计农园，让农园富有美感和生态价值。它的出现缘于人们在种地和维护城市绿化之间的矛盾。

　　The edible landscape is not so much a new thing as it is a revival of traditional culture. It is nothing new to plant crops and decorative plants together. The skills of edible landscapes had been used in the gardens of Babylon and ancient Egypt. As early as the 10th century, Benedictine monks had planted herbs on the periphery of rose gardens. It was not until the Renaissance that people began to consciously plant crops and purely decorative plants separately. Edible landscapes emphasize creativity, and a well-designed vegetable garden can be stunning! With fruit trees, vegetables, herbs, especially edible flowering plants, the edible landscape can make cities more beautiful and "delicious". The remaking of urban landscapes and the farm work in urban residential areas have long been nothing new to some Western countries. From community vegetable gardens to school farms, all those things enable city dwellers to enjoy the beauty of living a rural life. Moreover, with the increasing popularity of Permaculture's sustainable design concept, it has been a trend to transform one's own originally dull and uninteresting lawn into a beautiful and delicious edible landscape. In the 1980s, garden designer and environmentalist Robert Kourik invented this interesting term "edible landscape", which represents the integration of the landscape design and agricultural production. Subsequently, Rosalind Creasy published *The Complete Book of Edible Landscaping* in 1982, bringing this concept to the mainstream landscape design circle. The edible landscape is not simply the farm work, but designing a farm in the way of designing ecological gardens to make the farm rich in beauty and ecologically valuable. Its emergence is attributed to the contradiction between farming and the maintenance of urban greening.

　　可食地景在一定程度上改变了传统的耕种形式，使人们以新的方式进行农业生产。从设计、建造到产出，全程无污染、无公害，其自身产生的生物肥料，可进一步作为生物有机肥覆盖土地，以替代各种农药，减少各环节出现的环境污染，同时促进城市农产品和景观环境的更新。可食地景除能满足传统意义上的观赏价值外，更能给人们带来收获的喜悦（图 10-15、图 10-16），从种子萌芽到瓜熟蒂落的自然循环，这是大自然给予生命的完整过程，能让人在精神层面受到自然熏陶，获得愉悦感受，更能满足人们对土地的精神依赖。可食地景的环保价值也是显而易见的。常见的绿化草坪不仅耗费大量水资源，也要施用化肥和农药去维护它单一的生态系统，修剪草坪更是耗费大量人力物力。可食地景这一景观新模式，将可食性植物融入居住区建设中，提高居民户外活动的丰富性，营造更加多样性的景观环境，提升城市居住环境与生态环境品质，为打造城市居民可持续生存环境提供新思路。

图 10-15　收获（袁惠燕　摄）
Fig. 10-15　Harvest（by Yuan Huiyan）

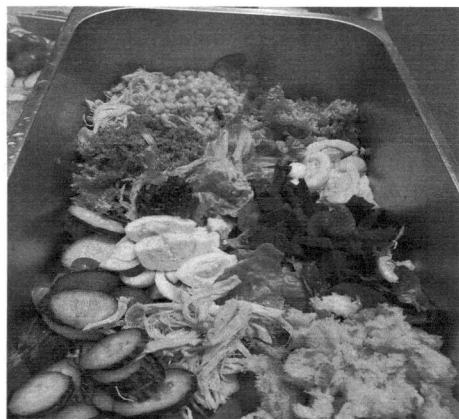

图 10-16　美味的蔬菜沙拉（袁惠燕　摄）
Fig. 10-16　Vegetable salad（by Yuan Huiyan）

The edible landscape has changed the traditional passive form to a certain extent, enabling people to carry out agricultural production in a new way. The whole process from design to construction to production is pollution-free, and the biological fertilizer produced by itself can be further used as bio-organic fertilizer to cover the land, replacing various pesticides, reducing environmental pollution in each process, and promoting the renewal of urban agricultural products and of the landscape environment. In addition to the traditional ornamental value, edible landscapes also bring people the joy of harvest (Fig. 10-15, Fig. 10-16). The natural cycle from seed germination to ripening of melons, the complete process of life given by nature, the nurture of nature and pleasant feelings on a spiritual level can better meet people's spiritual demand for the land and essential need to return to nature. The value of edible landscapes in environmental protection is obvious. The common green lawn not only consumes a lot of water resources, but also uses fertilizers and pesticides to maintain its single ecosystem; building a lawn takes much manpower and consumes many material resources. The new model of edible landscapes can integrate edible plants into the construction of residential areas, enrich residents' outdoor activities, create a more diverse landscape environment, improve the quality of urban living and ecological environments, and provide a new idea for building a sustainable living environment for urban residents.

10.4.1　可食地景常用植物种类
10.4.1　Commonly used plant species in edible landscapes

由于大部分可食用植物生长周期偏短、所需管养程度较高等原因，可食地景较常规的园林绿化造景而言，将面临更大挑战。营造与设计一个美观的可食用植物景观，需要对每种植物的特性都深入了解。比如紫菜苔，叶片是绿色，根茎却是紫红色，而开出的花是明亮的鲜黄色。

Due to factors such as a shorter growth cycle and the need for a higher level of maintenance of most edible plants, edible landscapes will face greater challenges than conventional landscapes created by greening. To create and design a beautiful edible plant landscape, we need to have a special understanding of the characteristics of each plant; for example, the leaves of *Brassica campestris* var. *purpuraria* are green, the rhizomes purplish red, and the flowers bright yellow.

10.4.1.1　一、二年生植物
10.4.1.1　Annual or biennial plants

这是指具有食用性和观赏价值的一类一、二年生植物。如各类生菜(图 10-17)、黄心苦菊、叶用甜菜(图 10-18)、瓜类(图 10-19)等，在可食地景设计中，既可美化环境，还可推动都市农业的发展。一些具有特殊景观效果的粮食作物，通过借助特定景观设计方法或与其他可食用植物及观赏植物种类搭配种植，也能打造出美丽的田园景观。在沈阳建筑大学新校园里，设计者用东北稻作为景观素材，设计了一片校园稻田，在稻田景观中，分布着一个个读书台，让稻香融入书声。在一个当代校园里，演绎了关于土地、人民、农耕文化的耕读故事，诠释了"白话"景观理念(俞孔坚，2007)。

图 10-17　生菜组合(袁惠燕　摄)
Fig. 10-17　Different Lettuces (by Yuan Huiyan)

Annual and biennial Plants with edible and ornamental value, such as all kinds of Lettuce (Fig. 10-17), Cichorium Endivia, Beets(Fig. 10-18), melons(Fig. 10-19) and so on, in the design of edible landscapes can beautify the environment and promote the development of urban agriculture. Some food crops with special landscape effects can be planted with the help of specific landscape design methods or planted in combination with other types of edible and ornamental plants to create beautiful pastoral landscapes. On the new campus of Shenyang Jianzhu University, designer used the Northeast rice as a landscape material to design a campus paddy field. Reading benches are dispersed in the paddy field landscape, blending the fragrance of rice into the sounds of reading aloud. On a modern campus, a story about the land, people, and farming culture is acted out, which interprets the concept of the "vernacular" landscape (Yu Kongjian, 2007).

图 10-18 甜菜、生菜、苦苣组合(袁惠燕 摄)

Fig. 10-18 Beets, Lettuce and Endives

(by Yuan Huiyan)

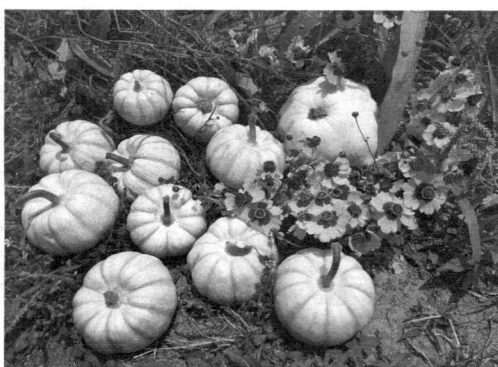

图 10-19 夏季迷你南瓜(袁惠燕 摄)

Fig. 10-19 Mini-pumpkins in summer

(by Yuan Huiyan)

10. 4. 1. 2 多年生植物
10. 4. 1. 2 Perennial plants

多年生植物是指一次播种或栽植，连续生长和采收在两年以上的植物。目前可食用的瓜、果和花卉等多年生植物在城市景观建设中的应用日趋广泛，如杧果、柿、银杏、杨梅等。可将这些瓜果植物设计成独具特色的瓜果廊；薄荷、萱草、百合、桔梗等芳香类多年生草本植物在可食地景的运用也很常见；利用可食药草营建景观，也是别具一格的园艺康健方法。

Perennial plants are those that have been sown or planted once but continuously grow and can be harvested for more than two years. At present, perennial plants such as the edible melons, fruits, flowers and plants are increasingly used in the urban landscape construction, for example, Mangos, Persimmons, Ginkgoes, and Red Bayberries. The flower stands, using edible melons, fruits and plants, can be designed into a unique melon and fruit gallery. It is also common to use aromatic perennial herbs such as Mint, Daylily, Lily, and Balloon Flower in edible landscapes. The edible herbs landscape also generates a unique horticultural therapy for health care.

10.4.2　植物配置与应用
10.4.2　The plants matching and application

可食地景设计时，在遵从科学的共生原理和生长周期的前提下，根据色彩搭配（图 10-20）、叶片渐变状态、菜苗密度等特点，合理播下蔬菜类、香草类等植物。考虑到景观的延续性，可在种植设计时，按一定比例配置同种不同成熟期的品种，如番茄的早中晚熟品种的搭配使用等。可食地景的协调发展，与其对环境条件的适用性有关，也与植物之间的相互作用密切相关。植物能通过体内分泌各种化感物质，作用于他种植物，以调节植物间的相互关系。如茄科植物明显抑制十字花科和蔷薇科植物，而芍药能分泌促进牡丹生长的化感物质。因此在可食地景设计时需要充分考虑植物之间的他感作用。

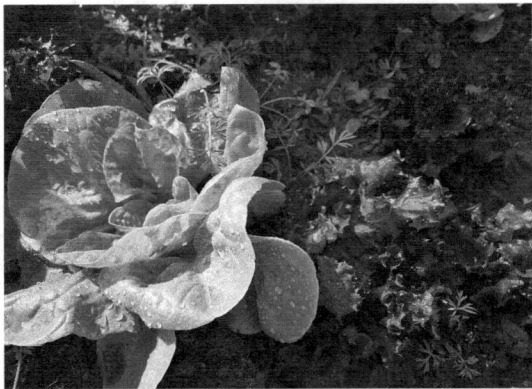

图 10-20　红配绿（袁惠燕　摄）
Fig. 10-20　Red and green leaves
（by Yuan Huiyan）

With the compliance with the scientific symbiosis principle and growth cycle as the premise, plants such as vegetables, herbs and so on should be sowed reasonably according to the characteristics of color matching (Fig. 10-20), gradual changes of leaves, seedling density and so on. In the planning and design of the edible landscape, to ensure the sustainability of the landscape, varieties of different maturity can be arranged in a certain proportion. An example of this is growing early, medium or late maturing varieties of tomatoes in combination. The coordinated development of edible landscapes is related to its applicability to environmental conditions, also closely related to the interplay of plants. Plants can act on other plants through various allelochemicals secreted in their bodies to regulate the interrelationships among plants; for example, Solanaceae plants significantly inhibit the growth of Cruciferae and Rosaceae plants, and Chinese herbaceous peonies can secrete allelochemicals that promote the growth of Peonies. Therefore, in the design of edible landscapes, it is necessary to fully consider allelopathy of plants.

10.4.2.1　种植制度
10.4.2.1　The planting system

合理的种植制度，有利于土地、阳光和空气、能源、水等各种资源的最有效利用，取得一定条件下植物生产的最佳社会、经济和环境效益，并能可持续地发展生产。目前可食地景采用的主要种植制度有轮作、单作、间作、混作、套作、立体种植等。

A reasonable planting system is conducive to the most effective use of various resources such as land, sunlight, air, energy, and water, and can realize the best social, economic, and environmental benefits of plant production under the conditions of the time, and sustainably boost production. The main planting systems currently available for edible landscapes include the crop rotation, single cropping, intercropping, mixed cropping, relay intercropping, and multi-story cropping.

(1) 轮作
(1) The crop rotation

轮作是在同一块田地上，有顺序地在季节间或年间，轮换种植不同的作物或复种组合的一种种植方式。轮作是用地养地相结合的一种生物学措施。

The crop rotation is a planting method in which different crops or multiple cropping combinations are sequentially planted on the same field across a sequence of seasons or years. The method is a biological measure that combines land use and land cultivation.

(2) 单作
(2) The single cropping

单作指在同一块田地上种植一种作物的种植方式。如每年热门景区——江西婺源油菜田就是一种比较典型的单作模式，主要欣赏植物的群体美。

The single cropping refers to the planting method of planting one type of crop on the same field, for example, the year's popular rape blossom fields in Wuyuan County, Jiangxi Province have adopted the typical single-cropping mode. It is the collective beauty of plants that people mainly appreciate.

(3) 间作
(3) Intercropping

间作指在同一田地上于同一生长期内，分行或分带相间种植两种或两种以上作物的种植方式。间作可提高土地利用率，由间作形成的植物复合群体，可增加对阳光的截取与吸收，减少光能的浪费；同时，两种作物间作还可产生互补作用。因此可食地景设计时对株形、高度、生育期稍有差异的植物进行合理搭配，有助于提高间作效果。

Intercropping refers to the planting method of planting two or more crops in alternate rows or strips on the same field in the same growth period. Intercropping can increase land use efficiency, and the plant community formed by intercropping can increase the interception and absorption of sunlight and reduce the waste of light energy; at the same time, the intercropping of two crops can generate complementary effects. Therefore, in the design of edible landscapes, a reasonable combination of plants with different heights and slightly different growth periods can help improve the effect of intercropping.

(4) 混作
(4) The mixed cropping

混作指将两种或两种以上生育季节相近的植物，按一定比例混合，种在同一块土地上的种植方式。如一年生羽扇豆或豌豆与燕麦混作，多年生三叶草与黑麦草混作。

The mixed cropping refers to the planting method of mixing two or more varieties of plants with similar growing seasons in a certain proportion on the same field. For example, the annual lupin or peas can be planted with oats in a mixed way, and the perennial clover with ryegrass.

（5）套作
（5）The relay intercropping

套作是指在前季植物生长后期的株、行或畦间，播种或栽植后季植物的种植方式。套作的两种或两种以上植物的共生期只占生育期的一小部分时间，是一种解决前后季植物间季节矛盾的复种方式。在可食地景上使用这种种植制度可保证一年四季有景可赏。

The relay intercropping describes a cropping pattern in which late-season crops are planted in a mixed way or in alternate rows or strips at a late stage of growth of early-season crops. The symbiotic period of two or more plants intercropped only accounts for a small part of the growth period, which is a way of multiple cropping to solve the seasonal contradiction between the late-season and early-season plants. The use of this planting system in edible landscapes can ensure that there is scenery to enjoy throughout the year.

（6）立体种植
（6）The multi-story planting

立体种植就是指充分利用立体空间的一种种植方式。现代立体种植可应用于公园绿化、社区或室内外装饰，在家庭室内、阳台、屋顶也可进行小型或微型的立体种植。

The multi-story planting refers to the planting method that makes full use of three-dimensional space. The modern multi-story planting can be used in park greening, community decoration or indoor and outdoor decoration, and small or micro multi-story planting can also be carried out in the interior, balcony, and roof of homes.

10.4.2.2　定植时间和密度
10.4.2.2　The planting time and density

为了缩短占地时间，提高土地利用率，增加观赏性，大多数可食地景植物通过育苗定植。适宜的定植或播种期，应根据当地的气候条件、植物的种类、品种及栽培目的来确定。一般落叶的果树、花卉在秋季植株落叶后或春季发芽前定植，蔬菜和草本花卉定植时期变化较大，可根据需要定植，但以春秋为主。

In order to shorten the land occupation time, increase the utilization ratio of the land, and improve the ornamental value, the planting of most plants for edible landscapes are done through seedlings cultivation. The appropriate planting or sowing period should be determined according to the local climatic conditions, types and varieties of plants and cultivation purposes. Generally, deciduous fruit trees, flowers and plants are planted in autumn when the plants' leaves have fallen or before germination in spring. The planting period of vegetables and herbaceous flowers and plants varies greatly. They can be planted at any time or according to needs, but mainly in spring and autumn.

定植密度也需要考虑。定植密度不仅影响产量和品质，更重要的是对景观效果的影响。合理密植，能最大程度地利用光热和土地资源，尽早达到设计预期的景观效果。定植密度受植株种类品种、当地气候和土壤条件、栽培方式和技术水平等因素的影响。

The plant density during planting is also an important issue that needs to be considered. The planting density not only affects the yield and quality, more importantly, but also impacts the landscape effect. The reasonable plant density can maximize the utilization of light, heat and land resources, and achieve the designed landscape effect as soon as possible. The plant density is affected by factors such as plant species, local climates and soil conditions, cultivation methods and engineering levels.

10.4.2.3　可食地景种植形式
10.4.2.3　The planting forms of edible landscapes

可食地景一般采用种植池的形式布局，种植池可以有多种形式。如可以高出地坪，由砖、石材或木板来框定边界，这种形式能使可食地景花园变得整洁、干净，防止松土时泥土撒落一地。也可以与地面齐平，由路径或砖石分割界限。还可以通过绿篱，框定种植池边界，在绿篱池中种植可食用的瓜果蔬菜。

Edible landscapes are generally arranged in the form of planting beds. Planting beds are of many forms, which can be higher than the ground and bounded by bricks, stones or wood planks. In this way, edible landscape gardens can be neat and clean, and the soil can be prevented from being scattered on the ground when the soil is loosened. Planting beds can also be level with the ground, divided by paths, bricks or stones. The edible landscape garden can become more beautiful by using hedgerows to form the boundary of planting beds and by planting edible fruits and vegetables in the hedgerow-lined planting beds.

可食地景植物的种植方式有以下几种：

The ways of planting the plants for edible landscapes can refer to the following layouts：

①规则长条形　预留种植时所需的道路，其他区域采用长条形的布局，规则分布，保持整齐感，每一条种植一种植物，形成线条感。

①The regular long strip　The roads needed during planting should be reserved, and the other areas should adopt the long-strip layout with regular distribution to maintain a sense of order, with one type of plant planted in each strip to form lines.

②规则方形　种植池的形状是正方形，阵列排布在场地中，每个方格种植一种植

物，方便后期管理。

②The regular square　The planting beds are square, arranged in rows in the garden, and in each square planted a type of plant, making it convenient to manage later.

③放射式　有一个中心，向四周引出道路，在道路之间设置种植操作区域。

③The radial concentric pattern　The roads radiate from a center, with planting operation areas set up between the roads.

此外，还可以将以上几种方式结合，形成几何式构图，将水景、花池与植物种植池结合，共组多样式景观。

In addition, the above several layouts can be combined to form geometric composition; waterscapes, flower beds and planting beds can be combined to form diverse landscapes.

10.4.2.4　可食地景四季管理养护
10.4.2.4　The management of edible landscapes in four seasons

可食地景倡导一种精细化的城市园林管理方式。日常管养工作主要围绕除草、施肥、灌溉进行。这些工作对熟悉有机种植的园丁而言并非难事，无须额外投入人力物力。可食地景是一项极具参与性和可操作性的项目，丝毫不逊于一般的装饰花园。简单易学的种植技术让居民轻松参与到可食地景的建设之中，分享劳动创造的果实。春季是播种的季节，整地施肥、定植植物、灌溉除草，可以结合不同的主题开展亲子活动；夏季植物旺盛生长、招蜂引蝶，令人心旷神怡，但该时节雨水较多，病虫害增加，要采取相应的预防措施；秋季是收获的季节，瓜果成熟，可以展开丰富的采摘类科普活动，同时要注意补种的速度，以延长可食地景的观赏性；冬季可种植露天越冬的植物，由于气温相对较低，病虫害少，灌溉的频率也要降低。另外，还需准备一处育苗区，便于及时更换品种，使可食地景一直保持"你退我进"的变化状态。清理下来的蔬菜又可作绿肥。可食地景不仅仅为人们提供新鲜安全的食物，还丰富了城市绿地的种类，是传统农耕文明的传承，更成为城市生物多样性的样本。图 10-21 至图 10-24 是苏州大学学生设计建造可食地景，以及在校园中享用可食地景丰硕成果的过程。

图 10-21　放线定植（袁惠燕　摄）
Fig. 10-21　Setting out and planting
（by Yuan Huiyan）

图 10-22　设计图落地（袁惠燕　摄）

Fig. 10-22　Completion based on design drawing

（by Yuan Huiyan）

图 10-23　享受劳动的成果（袁惠燕　摄）

Fig. 10-23　Enjoying the fruits of labor

（by Yuan Huiyan）

The concept of the edible landscape advocates a refined urban garden management method. The daily management work mainly revolves around weeding, fertilization, and irrigation. These tasks are not difficult for gardeners who are already familiar with organic planting, so additional input of manpower and material resources are not needed. The edible landscape is a very participatory and operatable program, no inferior to the general ornamental garden. The easy-to-learn planting skills allow everyone to easily participate in the construction of edible landscapes and share

图 10-24　与可食地景融为一体的雨水花园
（袁惠燕　摄）

Fig. 10-24　Rain garden with edible landscape

（by Yuan Huiyan）

the fruits of labor. Spring is the season for sowing, and soil preparation, fertilization, planting plants, irrigation and weeding can be combined with different themes to carry out parent-child activities; in summer, plants grow vigorously, attracting bees and butterflies, which is refreshing and comfortable, but at this time there is more rain and more plant diseases and insect pests, so it is necessary to take corresponding preventive measures; autumn is the harvest season when the fruits are mature, so a variety of scientific activities revolving around picking can be carried out; attention should be paid to the speed of replanting to extend the ornamental value of edible landscapes; in winter when the temperature is relatively low and when there are less pests and diseases, plants that can overwinter in the open air should be planted, and the frequency of irrigation should be reduced. In addition, it is necessary to prepare an area for seedling cultivation to facilitate timely replacement of varieties, so that the edible landscape can be maintained in a changing state of "dynamic balance". The vegetable remnants can be used as green manure. Edible landscapes not only provide people with fresh and safe food, but also enrich the types of urban green space. They have inherited the traditional farming civilization and become a sample of urban biodiversity. Fig. 10-21 to Fig. 10-24 show the process of students in Soochow University designing and building edible landscapes and enjoying the fruitful results of edible landscapes on their campus.

自人类文明诞生以来，有生产性的景观便与人的生活栖居紧密不分。在过去的二百年时间里，由于现代城市的发展，这种关系正在因为城市的单一发展模式被一点点削弱甚至临近瓦解。城市里自然气息的萌动被大量掩盖。而可食地景这种产出式的城市景观新方式，就是去触发和支持这样的可能性，让城市自然能量得到释放和拓展。产出式的城市景观新方式，可以让城市季节性地传达出一种活力，这种活力来自种植植物的变换，以及它们生长过程中呈现的变化，从播种到收获，甚至是需要直接暴露的土地休种期，其间所散发出的气味、声音和场景，这些都能让整个环境更加容易被人所理解和阐释。可食地景这种产出式的城市景观方式，可以在很大程度上为人们重新组织生活。

Since the birth of human civilization, productive landscapes have been inseparable from human life and habitation. In the past 200 years, due to the development of modern cities, the positive relationship that has been maintained over the centuries is being weakened bit by bit, even close to collapse, due to the single development model of cities. The natural atmosphere sprouting in the city is largely covered up. However, the edible landscape, a new mode of productive urban landscapes, is to trigger and support such possibilities and to release and expand the energy of urban nature. The new mode of the productive urban landscape enables cities to seasonally convey a kind of vitality, and this vitality comes from the changes of planted vegetation and in their growth process. The smells, sounds, and scenes from sowing to harvesting, and even during the fallow periods of the land that needs to be directly exposed can make it easier to understand and interpret this environment. The edible landscape, as a productive urban landscape, can re-organize life to a large extent.

思考题

1. 简述癌症花园的特点。

2. 简述上肢残障者和下肢残障者康复花园设计的共性和区别。

3. 从新加坡康复花园发展的历程和两个案例解析，谈谈你对当前我国普及康复花园循证设计的想法。

4. 请阐述可食地景常用的植物种类及其种植制度。

Questions

1. Please describe the characteristics of cancer garden.

2. Please describe briefly the similarities and differences in the design of healing gardens for people with upper and lower limb disabilities.

3. From the development of Singapore healing garden and the analysis of two cases, please talk about your idea of popularizing evidence-based design of healing gardens in China.

4. Please introduce plant species commonly used in edible landscapes and their planting system.

参考文献

References

曹林娣，2021. 听香深处——魅力[M]. 北京：中国电力出版社.

崔玖，林少雯，2012. 情绪的"处方"：花精[M]. 长沙：中南出版传媒集团，湖南人民出版社.

戴维·坎普，2016. 康复花园[M]. 潘潇潇，译. 桂林：广西师范大学出版社.

代星，2018. 基于园艺活动干预的幼儿记忆力比较研究[D]. 昆明：云南农业大学.

郭庭鸿，董靓，孙钦花，2015. 设计与实证：康复景观的循证设计方法探析[J]. 风景园林，22（9）：106-112.

郭文场，刘颖，段维和，等，1989. 中国药用花卉[M]. 北京：学术书刊出版社.

贾建平，2015. 中国痴呆与认知障碍诊治指南[M]. 北京：人民卫生出版社.

贾君兰，2020. 基于节气规律的适老型康复花园循证设计[D]. 苏州：苏州大学.

克莱尔·库珀·马库斯，2009. 康复花园[J]. 中国园林，25（7）：1-6.

克莱尔·库珀·马库斯，卡洛琳·弗朗西斯，2001. 人性场所——城市开放空间设计导则[M]. 俞孔坚，孙鹏，王志芳，等，译. 北京：中国建筑工业出版社.

理查德. 洛夫，2010. 林间最后的小孩[M]. 长沙：湖南科学技术出版社.

李卿，2013. 森林医学[M]. 北京：科学出版社.

李树华，2016. 中国特色园艺疗法体系建立的机遇与展望[C]∥2015 李树华. 中国园艺疗法研究与实践论文集. 北京：中国林业出版社：49-58.

李树华，2000. 尽早建立具有中国特色的园艺疗法学科体系（下）[J]. 中国园林，16（4）：32-34.

李先源，2018. 观赏植物分类学[M]. 北京：科学出版社.

梁友信，1987. 医学模式的转变与卫生毒理学[J]. 卫生毒理学杂志，1（1）：3-7.

刘滨谊，等，2016. 人居环境研究方法论与应用[M]. 北京：中国建筑工业出版社.

刘滨谊，2022. 走向景观感应：景观感知及视觉评价的传承发展[J]. 风景园林，29（9）：12-17.

刘玉龙，2006. 中国近现代医疗建筑的演进——一种人本主义趋势[D]. 北京：清华大学.

刘小勤，张衍泽，张承良，2005. 花卉栽培与药用[M]. 武汉：湖北科学技术出版社.

萝赛，2004. 花朵的秘密生命[M]. 钟友珊，译. 桂林：广西师范大学出版社.

美国园艺康健协会. 美国园艺康健协会官方网站[OL].［2017-03-20］. http://www.ahta.org/content.cfm/id/faq.

钱宏，张健，赵静超，2022. 世界上已知维管植物有多少种？基于多个全球植物数据库的整合[J]. 生物多样性，30（7）：22254.

孙旻恺，五岛圣子，浜野裕，2018. 日本以农村生活为蓝本的长崎园艺疗法对改善失智老人生活质量效果初探[C]. 2018 中国园艺疗法研究与实践论文集：75-82.

苏谦，辛自强，2010. 恢复性环境研究：理论、方法与进展[J]. 心理科学进展（18）：177-184.

苏州市园林管理局，2000. 苏州古典园林：汉英对照[M]. 上海：上海三联书店.

徐海滨，1996. 赏花指南[M]. 北京：中国农业出版社.

杨欢，刘滨谊，帕特里克·A·米勒，2009. 传统中医理论在康健花园设计中的应用[J]. 中国园林，25(7)：13-18.

俞孔坚，2007. 城市里的丰产稻田——沈阳建筑大学稻田校园设计[J]. 园林(9)：18-19.

张春晓，1996. 花卉栽培及药用[M]. 成都：四川科学技术出版社.

翟俊，2018. 景观都市主义的理论与方法[M]. 北京：中国建筑工业出版社.

郑丽，DERRICK STOWELL，李梦春，等，2016. 美国园艺疗法概况及行业术语最新解释[C]//李树华. 2015 中国园艺疗法研究与实践论文集. 北京：中国林业出版社：37-41.

郑丽，2020. 康健园艺与康复花园[M]. 苏州：苏州大学出版社.

周武忠，陈筱燕，1999. 花与中国文化[M]. 北京：中国农业出版社.

American Horticultural Therapy Association. Definitions and Positions［EB/OL］.（2019-2-23）.［2018-2-28］. http：//www. ahta. org/sites/default/files/DefinitionsandPositions. pdf

ARMSTRONG D，2000. A community diabetes education and gardening project to improve diabetes care in a northwest American Indian tribe[J]. Diabetes Educator，26(1)：113-120.

BARLEY E A，ROBINSON S. ，SIKORSKI J，2012. Primarycare based participatory rehabilitation：Users' views of a horticultural and arts project[J]. British Journal of General Practice，62：127-134.

BLAIR C K，MADAN-SWAIN A，LOCHER J L，2013. Harvest for health gardening intervention feasibility study in cancersurvivors[J]. Acta Oncologica，52(6)：1110-1118.

DIETTE G B，LECHTZIN N，HAPONIK E，et al. ，2003. Distraction therapy with nature sights and sounds reduces pain during flexible bronchoscopy：a complementary approach to routine analgesia[J]. Chest，123(3)：941-948.

ELSADEK M，SUN M，SUGIYAMA R，FUJII E，2019. Cross-cultural comparison of physiological and psychological responses to different garden styles[J]. Urban Forestry & Urban Greening，38：74-83.

ELSADEK M，LIU B，LIAN Z，2019. Green fagades：their contrib. stress recovery and well-being in high-density cities［J］. Urban Forestry & Urban Greening，46：126446.

EVA M SELHUB，ALAN C LOGAN，2012. Your Brain on Nature[M]. Ontario：John Willy & Sons Canada，Ltd.

FINNEMA EVELYN，DRÖES ROSE-MARIE，RIBBE MIEL，2000. The effects of emotion-oriented approaches in the care for persons suffering from dementia：a review of the literature［J］. International Journal of Geriatric Psychiatry. 15(2)：141-161.

GAYLE SOUTER-BROWN，2014. Landscape and Urban Design for Health and Well-Being[M]. New York：Routledge，Taylor & Francis Group.

GEORGE L ENGEL，1977. Need for a new medical model：a challenge for biomedicine[J]. Science，196(4286)：129-136.

GIGLIOTTI C M，JARROTT S E，2005. Effects of horticulture therapy on engagement and affect[J]. Canadian Journal on Aging-revue Canadienne du Vieillissement，24(4)：367-377.

HERZOG T R，BLACK A M，FOUNTAINE K A，et al. ，1997. Reflection and attentional recovery as distinctive benefits of restorative environments[J]. Journal of Environmental Psychology，17(2)：165-170.

IGNATIEVA M. ETC，2008. How to Out Nature into Our Neighbourhoods：Application of Low Impact Urban Design and Development Priciples，with a Biodiversity Focus，for New Zealand Developers and homeowners[M]. Lincoln：Mana Whenua Press.

I KEI H. ，SONG C. ，MIYAZAKI Y，2015. Physiological effect of olfactory stimulation by Hinoki cypress（Chamaecyparis obtusa）leaf oil[J]. Physiol. Anthropol，34：44.

IRVINE KN, WARBER SL, 2002. Greening healthcare: practicing as if the natural environment really mattered[J]. Altern Ther Health Med. , 8(5): 76-83.

KAPLAN R, KAPLAN S, 1989. The Experience of Nature: a Psychological Perspective [M]. New York: Cambridge University Press.

KAPLAN R, KAPLAN S, BROWN T, 1989. Environmental preference: a comparison of four domains of predictors[J]. Environment & Behavior, 21(5): 509-530.

KAPLAN S, 1995. The restorative benefits of nature: Toward an integrative framework[J]. Journal of Environmental Psychology, 15(3): 169-182.

KIMA E, YOUNG B, PARK SH, 2012. Horticultural therapy Program for the Improvement of Attention and Sociality in Children with Intallectual Disabilities[J]. Japanese Journal of Clinical Oncology, 22(3): 320-324.

LAUMANN K, GARLING T, STORMARK KM, 2003. Selective attention and heart rate responses to natural and urban environments[J]. Journal of environmental Psychology, 25(2): 125-134.

MAN D, OLCHAWA R, 2018. Brain biophysics: perception, consciousness, creativity. Brain Computer Interface (BCI) [C]. In Biomedical Engineering and Neuroscience: Proceedings of the 3rd International Scientific Conference on Brain-Computer Interfaces, BCI 2018. Opole: Springer International Publishing.

MATSUBARA E, FUKAGAWA M, OKAMOTO T, et al. , 2011. The essential oil of Abies sibirica (Pinaceae) reduces arousal levels after visual display terminal work[J]. Flavour et Fragrance Journal, 26(3): 204-210.

MINKAI SUN, KARL HERRUP, BERTRAM SHI, et al. , 2018. Changes in visual interaction: Viewing a Japanese garden directly, through glass or as a projected image[J]. Journal of Environmental Psychology, 60: 116-121.

NAKAU M, IMANISHI J, IMANISHI J, et al. , 2013. Spiritual care of cancer patients by integrated medicine in urban green space: a pilot study[J]. Explore, 9(2): 87-90.

NG K, SIA A, NG M, et al. , 2018. Effects of horticultural therapy on Asian older adults: A randomized controlled trial[J]. International Journal of Environmental Research and Public Health, 15(8): 1705. https: //doi. org/10. 3390/ijerph15081705

RAPPE E, TOPO P, 2007. Contact with outdoor greenery can support competence among people with dementia[J]. Journal of Housing for the Elderly, 21: 229-248.

SAHLIM E, AHLBORG JR G, MATYSZCZYK JV, et al. , 2014. Nature-based stress management course for individuals at risk of adverse health effects from work-related stress-effects on stress related symptoms, workability and sick leave [J]. International Journal of Environmental Research and Public Health, 11(6): 6586-6611.

SHANNON E JARROTT, 2010. Comparing responses to horticultural-based and traditional activities in dementia care programs[J]. American Journal of Alzheimer's Disease & other Dementias, 25(8): 657-665.

SIMSON S P, M C STRAUS, 2003. Horticulture as Therapy: Principles and Practice[M]. New York: The Haworth Press.

SODERBACK I, SODERSTROM M, SCHALANDER E, 2004. Horticultural therapy: The 'healing garden' and gardening in rehabilitation measures at Danderyd Hospital Rehabilitation Clinic, Sweden[J]. Pediatric Rehabilitation, 7: 245-260.

SOGA M, GASTON K J, YAMAURA Y, 2017. Gardening is beneficial for health: A meta-analysis [J]. Preventive Medicine Reports, 5: 92-99.

STICHLER J F, HAMILTON D K, 2008. Evidence-based Design: What is It? [J]. Health Environments Research and Design, 1(2): 3-4.

SZOFRAN J, 1998. Case studies part 2: physical disability. Ethel: Diagnosis chronic degenerative arthritis, bilateral total knee replacement[J]. J. Hortic, 9: 31-32.

TAYLOR A F, KUO F E, SULLIVAN W C, 2001. Coping with ADD: The surprising connection to green play settings[J]. Environment and Behavior, 33: 54-77.

ULRICH R S, 1983. Aesthetic and Affective Response to Natural Environment[M]. New York: Plenum Press: 85-125.

ULRICH R S, 1984. View through a window may influence recovery from surgery[J]. Science, 224 (4647): 420-421.

ULRICH R S, 1999. Effects of Gardens on Health Outcomes: Theory and Research[M]. New York: Wiley: 27-86.

ULRICH R S, PARSONS R, 1992. Influences of Passive Experiences with Plants on Individual Well-being and Health[M]. OR: Timber Press.

ULRICH R S, SIMONS. 1986. Recovery from Stress During Exposure to Everyday Outdoor Environments. In J. Wineman, R. Bames, and C. Zimring(eds.), The Costs of Not Knowing: Proceedings of the Seventeenth Annual Conference of the Environmental Design Research Association [C]. Washington, DC: Environmental Design Research Association.

VERRA M L, ANGST F, BECK T, et al., 2012. Horticultural therapy for patients with chronic musculoskeletal pain: results of a pilot study[J]. Alternative Therapies in Health and Medicine, 18(2): 44-50.

VIETS E, 2009. Lessons from evidence-based medicine: what healthcare designers can learn from the medical field [J]. Health Environments Research and Design Journal, 2(2): 73-87.

WEISMAN GD, COHEN U, RAY K, et al., 1991. Architectural Planning and Design for Dementia Care Units. Baltimore: The Johns Hopkins University Press.

WHITEHOUSE S, VARNI J W, SEID M, et al., 2001. Evaluating a children's hospital garden environment: Utilization and consumer satisfaction[J]. Journal of Environmental Psychology 21, 301-314.

WICHROWSKI M, CHAMBERS NK, CICCANTELLI LM, 1998. Stroke, Spinal Cord and Physical Disabilities in Horticulture Therapy Practices[M]. New York: Food Products Press: 71-104.

WICHROWSKI M, WHITESON J, HAAS F, et al., 2005. Effects of horticultural therapy on mood and heart rate in patients participating in an inpatient cardiopulmonary[J]. Journal of Cardiopulmonary Rehabilitation, 25(5): 270-274. https://www.who.int/zh/about/governance/constitution

安川緑, 千葉茂, 伊藤喜久, 等, 2005. 認知症高齢者に対する園藝療法の有効性に関する研究[J]. 人間・植物関係學會雜誌 (5): 20-21.

長倉壽子, 森本惠美, 時政昭次, 等, 2009. 小集團活動が中等度認知症に有する高齢者のBPSDに及ぼす影響[J]. 老年精神醫學雜誌, 20(12): 1401-1408.

黒田利香, 小西美智子, 寺岡佐和, 等, 2001. 特別養護老人ホームにおけるアクティビティケアとしての園藝活動の効果[J]. 廣島大學保健學ジャーナル, 1(1): 49-53.

鈴木みずえ, 磯和勅子, 2004. 大きな可能性を秘める高齢者のアクティビティケア[J]. コミュニティケア 6(3): 68-71.

梅田みちる，杉美惠，竹重都子，等，2001. 痴呆性高齢者に対する園藝療法の有用性の検討：療養型病床群における試み[J]. 日本看護學會論文集老人看護(32)：140-142.

清水邦義，孫旻愷，鬆本清，等，2018. 睡眠誘導効果をもつ(-)-醋酸ボルニルの睡眠促進効果ならびに肌質改善効果[J]. フレグランスジャーナル：香妝品科學研究開發專門志，46(2)：24-30.

杉原式穂，2011. 認知症高齢者に対する園藝療法[J]. 老年精神醫學雜誌，22(1)：22-26.

寺岡佐和，原田春美，2003. 施設入居痴呆高齢者 QOL 向上に寄與する園藝療法とその評價方法[J]. Quality Nursing, 9(7)：21-27.

孫旻愷，本傳晃義，羽賀榮理子，等，2019. スギの無垢材を内装に用いた室内空間が人の睡眠に及ぼす影響[J]. 木材工業，74(7)：266-271.

田崎史江，2005. 認知症における非藥物療法實踐レポート 最終回園藝療法の實驗と認知症高齢者への効果[J]. 介護リーダー，10(6)：105-114.

増谷順子，2010. グループホーム入居の認知症高齢者への園藝活動の試み[J]. 日本認知症ケア學會志，9(3)：552-563.

附 录
Appendixes

附录1　残障人士服务注意事项
Appendix 1　Notes for service of disabled persons

为残障人士服务
Serving People With Disabilities

作为一名园艺康健师，你在塑造残障人士的公众形象方面有着独特的地位。你所使用的文字和图片可以为残障人士创造积极的形象。

As a horticultural therapist, you are in a unique position to shape the public image of people with disabilities. The words and images you use can create a positive view of people with disabilities

残疾是指身体或精神上的缺陷，这种缺陷限制了一个人日常活动的一个或多个方面。我们都在生命中的某个时刻经历过损伤，但对于有永久性或长期残疾的人来说，这一损伤构成了他们生活中更重要的因素。虽然损伤不能被定义为残疾，但这种损伤确实严重影响了他们的生活。

Disability means a physical or mental impairment that restrict one or more aspects of a person's daily activity. We all experience disability at some point in our lives. But for people with permanent or long term disabilities, impairment is a more significant factor in their lives. While people with disabilities are not defined by impairment, impairment does play a role in shaping their lifestyles.

大多数残障人士只是在狭窄的活动范围内存在困难，而不是在更广泛的社会、职业和认知行为方面。残障人士能够而且确实参与了生活的各个方面，包括工作、娱乐、恋爱和养育子女。残障人士与其他人一样，也能体验到积极的生活质量。他们希望得到尊重，与非残障人士一样受到平等对待。

Most people with disabilities have difficulties in just a narrow range of activity, not their wider scope of social, vocational and cognitive behavior. People with disabilities can and do participate in all aspects of life, including work, play, romance and parenting. People with disabilities experience a positive quality of life to the same degree as other people. People with disabilities want to be treated with respect and as equals with their non-disabled peers.

残疾是普遍存在的，包括任何背景、男性/女性和任何年龄的人。残疾源于先天缺陷、疾病或损伤的影响，而不是身体或精神上的状况，主要是由于居所的桎梏和人们世俗的偏见等不友好的环境，阻碍了残障人士本可以像正常人一样的生活。换言之，正是这些障碍才导致了"残疾"。

Disability is universal, encompassing people of all backgrounds, male/female and any age. Disabilities stem from impairments that are congenital or the residual effects of disease or injury. Handicaps, by contrast, are not physical or mental conditions. They are the architectural and attitudinal barriers that impede individuals trying to function in a non-friendly environment. In other words, a person is handicapped by a barrier or obstacle.

另一个问题是一些人对残障人士感到不舒服。当人们不熟悉残障人士的社会和医疗方面的情况时，他们常常对差异或未知怀有恐惧。有些人可能会对他们不理解的人或情况感到害怕或感觉受到威胁。这可能导致假设、刻板印象和偏见。当遇到残障人士时，最好的办法就是敞开心扉，愿意学习，不做任何假设。

Another issue is the discomfort some people feel around people with disabilities. The fear of differentness or the unknown is common when people are unfamiliar with the social and medical aspects of disability. Some people may feel frightened or threatened by a person or situation that they do not understand. This can lead to assumptions, stereotyping and prejudice. When meeting people with disabilities, the best thing to do is to be an open slate, be willing to learn and make no assumptions.

＊残障人士具有各种各样的个性特征。残疾不是衡量性格的标准。

＊People with disabilities possess the full range of personality traits. Disability is not a measure of character.

＊残疾不会传染。

＊Disability is not contagious.

＊大多数残障人士希望促进理解。如果你有关于残疾的问题，请礼貌询问，并且问题要与谈话有关。

*Most people with disabilities want to promote understanding. If you have questions about a disability, ask within polite boundaries and your question is relevant to the conversation.

*残障人士愿意多想自己的优点，不愿多想自己的缺点。

*People with disabilities would rather dwell on their strengths than their weaknesses.

除了一些人的不适和恐惧之外，残障人士还会遇到身体和环境形成的障碍，这些障碍会使日常生活变得复杂。例如，杂货店、停车设施和公共交通系统的设计，往往使日常工作变得困难。以下指导方针有助于创造一个更友好的环境：

In addition to some other people's discomfort and fear, people with disabilities encounter physical and situational barriers that can complicate everyday living. Grocery stores, parking facilities, and public transportation systems, for example are often designed in ways that make routine errands difficult. These guidelines can help create a friendlier environment：

*如果你认为某人需要帮助，就去问。问总是可以的！但假设是不对的。

*If you think someone needs help, ask. It's always okay to ask！It's not okay to assume.

*一旦开口，在提议被接受之前不要行动。有些人喜欢单独做事。

*Once you ask, don't move until your offer is accepted. Some people prefer to go it alone.

*询问如何提供帮助，然后按照对方的指示去做。

*Ask how to be of assistance, and then follow the person's instructions.

*当安排与残障人士会面时，要确保地点方便可达。

*When arranging to meet someone with a disability, be sure the location is accessible.

*总是朝着你正在交谈的人说话，而不是朝着你的同伴或翻译说话。

*Always speak to the person you're addressing, not a companion or interpreter.

*说话或写作时，用词应该以人为本，如残障人士、脑瘫患者等。

*When speaking or writing, put the person first：person with a disability, person with cerebral palsy, etc.

态度和语言
Attitudes and Language

残障人士：
People with disabilities：

可以说身有残障，而不要说残障缠身、患有残疾，或者是残疾受害者。

Are living with their disabilities…. Are not suffering from, victims of, or afflicted by their disabilities.

可以说他们正在克服他们周围的建筑或人们的态度所造成的障碍……但不要说他们在克服他们的残疾。

Are overcoming the barriers that the architecture or attitudes of people surrounding them create···Are not overcoming their disabilities.

他们希望被描绘成找到其他方法来完成日常活动的人······不希望被描绘成勇敢或受折磨的人。

Want to be portrayed as individuals who find alternative means to accomplish everyday activities···Do not want to be portrayed as courageous or tortured.

他们更喜欢被称为"残障人士"。这强调的是人，而不是残疾。

Prefer to be called just that："people with disabilities."This emphasizes the person, not the disability.

通行说法	冒犯性语言
残障人士	残疾、残废
听力受损人士	聋子、哑巴
盲人或视力受损的人	瞎子
轮椅使用者	坐轮椅的
行动不便的人	瘸子、瘸腿的
有发育障碍，智力缺陷、认知障碍或精神发育迟滞	智障、弱智、傻子
有精神疾病的人	精神错乱、疯子、傻子、变态、精神分裂

Accepted Language	Offensive Language
People with disabilities	The handicapped, the disabled
People who are deaf or hard of hearing	The hearing impaired, deaf-mute, stone-deaf
People who are blind or visually impaired	The blind, the sightless
Wheelchair users	Confined to wheelchairs, wheelchair bound
People with mobility impairments	The crippled, the lame
People who have developmental Disabilities, cognitive disabilities, or mental retardation	The retarded, the mentally deficient
People with mental illness	The insane, crazy, mad, psycho, schizo

避免使用这些术语：

Avoid these terms

受到折磨	完全瞎了
瘸了、瘸子	又聋又哑；聋子哑巴
缺陷、有缺陷的	畸形
某某之家/失智老人之家	无法离家的
伤残者	
正常的(作为残疾的反义词)	
可怜的	可悲的、不幸的
得病的	受害者

被束缚在轮椅上的

Afflicted/afflicted with/afflicted by	Blind as a bat
Cripple/crip/crippled/the crippled/crippling	Deaf and dumb, deaf mute
Defect/defective	Deformed
Group home	Homebound
Invalid	
Normal (as the opposite of having a disability)	
Pitiful	Poor, unfortunate
Stricken	Victim
Wheelchair bound/confined to a wheelchair	

使用轮椅、拐杖或手杖的人
People who use wheelchairs, crutches, or canes

①以人为本。把有残疾的来访者视为一个人，而不是残障人士。尊重所有访客。

①People first. See the visitor who has a disability as a person, not as a disability. Treat all visitors with respect.

②直接与残障人士交谈，而不是与同伴或翻译交谈。

②Speak directly to the person who has a disability, not to a companion or an interpreter.

③要有耐心。残障人士说话或做事可能需要额外的时间。

③Be patient. It may take extra time for the person with a disability to say or do things.

④不要问关于残疾的私人问题，除非你很了解这个人，并且知道他们愿意谈论这个话题。

④Do not ask personal questions about a disability unless you know the person well and know that they are comfortable discussing the subject.

⑤除非在场的每个人都很熟悉，否则不要直呼其名。

⑤Do not use first names unless that familiarity is extended to everyone present.

⑥放松。当与视力或身体残疾的人交谈时，不要担心使用诸如"回头见"或"我得走了"之类的常用表达。

⑥Relax. Don't worry about using common expressions like, "see you later" or "I've got to be running along" when talking to people with visual or physical disabilities.

⑦不要认为残障人士需要帮助。介绍你自己是 CBG(Community-Based Group，社区服务团体)的一员，等到别人接受你的帮助后再提供帮助。

⑦Don't assume the person with a disability needs help. Introduce yourself as a member of CBG (Community-Based Group) and wait until help is accepted before giving it.

⑧把残障人士当作健康人对待。如果一个人有功能限制，并不意味着这个人生病了。有些残疾没有伴随的健康问题。

⑧Treat a person with a disability as a healthy person. If a person has a functional limitation that does not mean the individual is sick. Some disabilities have no accompanying health problems.

⑨未经允许，不要抚摸导盲犬或其他服务性动物。这些动物不是宠物。它们正在工作，必须时刻注意它们的同伴。

⑨Do not pet a guide dog or other service animal without asking permission. These animals are not pets. They are working and must stay attentive to their companions.

⑩一定要直接对来访者说话，而不是对同伴或随从说话。

⑩Be sure to speak directly to the visitor rather than to a companion or attendant.

⑪询问一个坐轮椅的来访者是否愿意被推。未经访客要求或许可，请勿推轮椅。

⑪Ask a visitor using a wheelchair if he or she wants to be pushed. Do not push the wheelchair without the visitor's request or permission.

⑫不要靠在或依附在别人的轮椅上，这是他/她个人空间的一部分。

⑫Do not lean on or hang onto a person's wheelchair. It is part of his/her personal space.

⑬坐着或蹲着，与坐在轮椅上的人处于同一视线水平。如果谈话持续了几分钟以上，那就给自己找张椅子。

⑬Sit or squat so that you are on the same eye level as the person in a wheelchair. If conversation continues for more than a few minutes, get a chair for yourself.

⑭允许使用轮椅、拐杖或手杖的游客将它们放在触手可及的地方。许多轮椅使用者可以把自己转移到椅子或汽车座椅上。有些人可以用拐杖或手杖短时间行走。

⑭Allow visitors who use wheelchairs, crutches, or canes to keep them within reach. Many wheel chair users can transfer themselves to chairs or car seats. Some can walk with crutches or canes for short periods of time.

⑮指路时，要考虑距离、天气状况和水平变化，如楼梯、路边或斜坡。

⑮When giving directions, consider distance, weather conditions, and level changes such as stairs, curbs, or inclines.

有视觉障碍的人

People with visual impairments

①介绍自己和其他可能和你在一起的人，用正常的音量和语调说话。永远不要离开一个盲人而不告诉他/她你要离开。

①Introduce yourself and any others who may be with you. Speak with your normal volume and tone of voice. Never walk away from a person who is blind without telling him/her you are leaving.

②在开始谈话时直接称呼来访者，这样他/她就知道你在和谁说话。

②Address the visitor directly when starting a conversation so he/she knows to whom you are speaking.

③要注意，嘈杂的环境可能会分散依赖听觉的人的注意力。

③Be aware that a noisy environment may be a distraction for a person who relies on hearing.

④未经允许不得抚摸导盲犬。导盲犬在工作，需要保持警惕。走在远离导盲犬的

一侧。

④Do not pet a guide dog without asking permission. The dog is working and needs to stay alert. Walk on the side of the person that is away from the dog.

⑤并非所有有视力障碍的人都阅读盲文。有些人阅读大字或使用录音机或其他特殊设备。

⑤Not all people with vision impairments read Braille. Some read large print or use tape recorders or other special equipment.

⑥询问来访者是否需要帮助。在给予帮助时，让对方挽着你的手臂，而不是你挽着他/她的手臂。用你正常说话的声音提醒对方任何台阶、路况变化、平整或狭窄的通道。使用诸如"在你的左边"和"在你的右边"之类的具体用语。

⑥Ask the visitor if he/she wants help. When giving assistance, allow the person to take your arm, rather than your taking his/her arm. Warn the person in your normal speaking voice of any steps, changes, in level, or narrow passageways. Use specifics such as "on your left" and "to your right".

⑦为了帮助安排座位，你应该征得来访者的同意，将他/她的手放在椅背或扶手上。

⑦To assist with seating, you should ask the visitor's permission to place his/her hand on the back or arm of the chair.

有语言障碍的人
People with speech difficulties

①如果你很难理解来访者所说的话，不要假装你听懂了。重复你所理解的内容，并注意对方的反应。

①If you have difficulty understanding something a visitor has said, do not pretend that you understand. Repeat as much as you have understood and be attentive to the person's response.

②要有耐心。不要纠正或替别人说话。允许额外的时间，并在需要的时候给予帮助。实际上，当你花时间去理解他/她在说什么的时候，每个说话有很大困难的人都会感激你。

②Be patient. Do not correct or speak for the person. Allow extra time and give help when needed. Virtually everyone with significant difficulty speaking appreciates when you take time to understand what he/she is saying.

③如果可能的话，问一些需要简短回答、点头或摇头的问题。

③If possible, ask questions that require short answers or a nod or shake of the head.

④如果情况允许，把你全部的注意力放在说话有困难的人身上。

④If the situation permits, give your complete attention to the person who has difficulty speaking.

⑤如果你对一个特定的回答有困难，请对方拼出来，写下来，或者重新表述。

⑤If you have difficulty with a particular response, ask the person to spell it, write it down, or rephrase it.

⑥语言障碍与智力无关。

⑥Speech impairments are not related to intelligence.

⑦大多数有语言障碍的人都能听到。大声或简单的词语并不更容易理解。

⑦Most people with speech impairments can hear. Loud or simple words are not easier to understand.

失聪的人或有听力障碍的人
People who are deaf or hard of hearing

①引起对方的注意。挥手以获得眼神交流。在紧急情况下，可能有必要轻拍对方的肩膀或摇灯。

①Get the person's attention. Wave your hand to gain eye contact. In an urgent situation, it may be necessary to tap the person's shoulder or flick the lights.

②直接对失聪或有听力障碍的来访者说话，而不是对同伴或翻译说话。

②Speak directly to the visitor who is deaf or hard of hearing, not to a companion or interpreter.

③助听器可能只是部分有效。

③Hearing aids can be just partially effective.

④即使声音被放大，对于听力受损的人来说，声音也可能是扭曲的。

④Even when amplified, sounds may seem distorted to someone with hearing loss.

⑤请记住，手语翻译是聋人和正常人之间沟通的桥梁。当译员在翻译时，不要向他/她询问建议或信息。

⑤Remember that a sign language interpreter is a communication bridge between deaf and hearing people. Do not ask the interpreter for advice or information while he/she is interpreting.

⑥说话清晰，语速正常。不要大声喊叫或夸大嘴唇动作。保持句子简短。

⑥Speak clearly and at a normal speed. Do not shout or exaggerate lip movements. Keep sentences short.

⑦语言要灵活。如果对方理解有困难，你可以重新措辞。如果困难仍然存在，试着把信息写下来。

⑦Be flexible in your language. If the person has difficulty understanding you rephrase your statement. If difficulty persists, try writing down the message.

⑧要有耐心。如果你不明白访客在说什么，不要猜测或假装你懂。

⑧Be patient. If you do not understand what the visitor is saying, do not guess or pretend that you do.

⑨提供一个清晰的视野，让光源照在脸上。说话时，手、食物或其他任何东西都不要放到嘴边。永远面对个人或团体。

⑨Provide a clear view of your face and keep the light source on it. Keep hands, food, or anything else away from your mouth when talking. Always face the person or group.

⑩做一个活泼的演讲者。使用与你的语气相匹配的面部表情，并使用手势和肢体动作来帮助沟通。

⑩Be a lively speaker. Use facial expressions that match your tone of voice and use gestures and body movements to aid communication.

精神疾病患者
People with mental illness

①请记住，"精神疾病"是一个通用术语，用来描述那些在思考、感觉和人际关系方面有严重障碍的人。

①Keep in mind that "mental illness" is a general term used to describe people who have severe disturbances in thinking, feeling, and relating.

②举止自然，对待来访者就像对待任何来访者一样。保持眼神交流，不要居高临下。一定要记住，谈话的水平要适合来访者或团体的年龄。

②Act naturally and treat the person just as you do to any visitors. Maintain eye contact and do not condescend. Always remember to talk at a level appropriate to the age of the visitor or group.

③如果访客似乎不知道园内的安全行为准则，请用准确的句子，清楚而简短地说明规则。

③If a visitor seems to be unaware of safe Garden conduct, clearly and briefly state the rules, using precise sentences.

④不要大喊大叫。安静、低沉的声音是最有效的。

④Do not shout. A quiet, low-pitched voice is the most effective.

⑤在花园这样刺激元素丰富的环境中，一个人可能会冲动地说话或做事，或者变得过度兴奋或沮丧。体谅每个人，但不允许有不恰当的行为。必要时向上级寻求帮助。

⑤In a stimulating environment like a garden, a person may say or do things impulsively or become overly excited or frustrated. Be considerate of any individual but do not allow inappropriate behavior. Seek assistance from a supervisor if necessary.

有认知障碍/智力障碍者
People with cognitive disabilities/mental retardation

①直接对来访者说话，而不是对同伴。

①Speak directly to the visitor, not to a companion.

②慢慢地、清晰地说。对于智力障碍或发育障碍的人来说，处理信息可能需要更长的时间和耐心。

②Speak slowly and distinctly. Processing information may take longer for a person who is mentally retarded or developmentally disabled. Be patient.

③展示可能比诉说更有效。

③Showing might be more effective than telling.

④告诉访问者该做什么，而不是不该做什么。

④Tell the visitor what to do instead of what not to do.

⑤让来访者感到舒适。保持不具威胁性的语调和面部表情。

⑤Help the visitor feel comfortable. Maintain a non-threatening tone of voice and facial expression.

⑥以对待成年人的方式对待智力障碍的成年访客。

⑥Treat the adult visitor who has mental retardation as an adult.

附录 2　中国传统十大名花药性（按中文名拼音排序）
Appendix 2　Medicinal properties of Chinese Top Traditional Famous Flowers（by pinyin alphabetic order according to the flowers' initial of Chinese names）

1. 杜鹃花

学名：*Rhododendron simsii*

别名：鹃花、映山红、满山红、红踯躅

科名：杜鹃花科

药用：

采集加工　花初放时采摘，夏季采叶，秋冬采根，晒干备用或鲜用。

性味　花、叶：甘、酸，平。根：酸、涩，温。有毒。

功效　花、叶：清热解毒，化痰止咳，止痒。根：祛风湿，活血化瘀，止血。

1. Azalea

Scientific name：*Rhododendron simsii*

Alias：Rhododendron；*Rhododendron dauricum*；Rhodora

Family name：Ericaceae

Collecting and processing for medicine：Flowers are picked when they start to bloom. Leaves are picked in summer, roots in autumn and winter, freshly used or dried for later use.

Property and flavor：The flower and the leaf are both sweet and sour in flavor, mild in nature. The root is sour and astringent in flavor, warm in nature. Poisonous.

Efficacy：The flower and the leaf can clear internal heat and remove toxicity, eliminate phlegm and stop cough, and relieve itch. The root can expel internal dampness, promote blood circulation to remove blood stasis, and stanch bleeding.

2. 桂花

学名：*Osmanthus fragrans*

别名：岩桂、木樨、丹桂、九里香

科名：木樨科

品种：'金桂'：花深黄色，香气最浓，花朵容易脱落。'银桂'：花白色微黄，香气较'金桂'淡，花朵较牢固。'丹桂'：花橙黄色，美丽，香气亦较浓。'四季'桂：花乳黄色，花期长，以秋花较盛。花香较前三种淡，且每次开花量少。

药用：

采集加工　花于秋季开时采收，阴干，撩去杂质，密闭贮藏，防止走失香气及受潮发霉。早春采籽（桂花中只有'四季'桂、'银桂'能结籽）。用温水浸泡后，晒干。

枝叶和根，四季可采。

　　性味　花、果、枝、叶：甘、辛、温。根：甘、微涩，平。

　　功效　花：散寒破结，化痰止咳。果：暖胃、平肝、散寒。枝、叶：温中散寒、暖胃止痛。根：祛风湿、散寒。

2. Sweet Osmanthus

Scientific name：*Osmanthus fragrans*

Alias：Cinnamomum Petrophilium；Sweet-scented Osmanthus；Aurantiacus；Murraya exotica

Family name：Oleaceae

Species：

Thunbergii：Flowers are deep yellow，with the strongest aroma，easy to fall off.

'Odoratus'：Flowers are tight，white but slightly yellow，with lighter fragrance than that of Thunbergii.

'Aurantiacus'：Flowers are orange-yellow with strong aroma，magnificently beautiful.

'Semperflorens'：Flowers are creamy yellow and have a long flowering period，with autumn flowers being more prosperous. The scent is lighter than the first three mentioned above，and the flowers are fewer during the flowering phase.

Collecting and processing for medicine：Flowers are picked when they bloom in autumn，dried in shade with impurities removed，and stored in seal to prevent damp mold and loss of fragrance. Seeds are picked in early spring（only 'Semperflorens' and 'Sweet Osmanthus' *nakai* can have seeds），and later soaked in warm water and dried. Branches and roots can be picked in all seasons.

Property and flavor：The flowers，fruit，branches and leaves are sweet and pungent in flavor，warm in nature. The root is sweet and slightly astringent in flavor，mild in nature.

Efficacy：The flower can dispel internal cold and eliminate coagulation，eliminate phlegm and stop coughs. The fruit can warm the stomach，calm the liver and dispel internal cold. The branches and leaves can warm the spleen and the stomach，dispel internal cold，and relieve pains. The root can be used to expel internal dampness and cold.

3. 荷花

　　学名：*Nelumbo nucifera*

　　别名：莲、芙蕖、水芙蓉、藕花、水华、水芸、菡萏

　　科名：睡莲科

　　药用：

　　采集加工　藕：秋、冬及初春采挖。莲子：9~10月待成熟时采收，剪下莲蓬，去壳皮，晒干。荷花：待花初放或含苞未放时采收，阴干备用，鲜用也可。荷叶：夏秋采之。藕粉：藕加工而得。

性味　生藕：甘，寒。熟藕：甘，温。莲子：甘，涩，平。荷花：苦，甘，温。荷叶：甘，微苦，平。藕粉：甘，咸，平。

功效　生藕：凉血散瘀。熟藕：补心益胃。莲子：益肾固精。荷花：祛湿消风。荷叶：清热解暑。藕粉：调中开胃。

3. Lotus

Scientific name：*Nelumbo nucifera*

Alias：Water Lotus；Water Bloom；Water Lily

Family name：Nymphaeaceae

Collecting and processing for medicine：Lotus roots can be dug out and collected in autumn, winter and early spring for fresh use or have them cooked. As to lotus seeds, they can be taken out freshly by cutting the lotus open with a sharp knife when the lotus are picked from September to October when they are ripe. Then skin of the seeds are peeled off and dried. The lotus flowers are collected when they first blossom or ready to bloom, then dried in the shade for later use or used fresh. The lotus leaves are picked in summer and autumn. Lotus powder is made by processing lotus roots.

Property and flavor：Raw lotus root：Sweet in flavor, cold in nature. Cooked lotus root：Sweet in flavor, warm in nature. Lotus seeds：Sweet and astringent in flavor, mild in nature. Lotus flower：Bitter and sweet in flavor, warm in nature. Lotus leaf：Sweet and slightly bitter in flavor, mild in nature. Lotus root starch：Sweet and salty in flavor, mild in nature.

Efficacy：Raw lotus root：Clear away the heat pathogen in the blood and disperse blood stasis. Cooked lotus root：Nourish the heart and benefit the stomach. Lotus seeds：Nourish the kidneys and enhance the kidneys' ability to store essence. Lotus flower：Eliminate dampness and expel wind-evil. Lotus leaf：Alleviate the pathogenic heat and summer-heat evils. Lotus root starch：Regulate the functions of the spleen and stomach to enhance appetite and promote digestion.

4. 菊花

学名：*Chrysanthemum morifolium*

别名：黄花、节花、秋菊、九花、菊华

科名：菊科

药用：

采集加工　药用菊花主要有三种："黄菊花"(产于浙江杭州的最好，所以称"杭菊花")；"白菊花"(产于安徽滁县的最好，所以又称"滁菊花")；"野菊花"。通常于农历九月花盛开时，选晴天露水干后或下午，采摘花头晒干。阴雨天亦可将花采下，用小火烘干，火不宜太旺，防止烘焦。

性味　甘苦，微寒。

功效　疏风散热，养肝明目，解疔疮毒。

4. Chrysanthemum

Scientific name：*Chrysanthemum morifolium*

Alias：Wreath Goldenrod；Autumn Chrysanthemum

Family name：Asteraceae

Collecting and processing for medicine：There are mainly three types of medicinal chrysanthemums："Yellow Chrysanthemum"（the best is known as "Hang Chrysanthemum", produced in Hangzhou，Zhejiang Province）；"White Chrysanthemum"（the best is known as "Chu Chrysanthemum"，produced in Chu County，Anhui Province）；and Wild Chrysanthemum. When flowers are in full blossom in September of the lunar calendar；the flower heads are picked and dried on a sunny day after the dew is dry or in the afternoon. The flowers can also be picked in rainy days and dried with low heat to prevent scorching.

Property and flavor：Both bitter and sweet，slightly cold in nature.

Efficacy：Expel internal dampness or heat，nourish liver，improve eyesight，and relieve poison of skin ulcer.

5. 兰花

学名：*Cymbidium*

别名：兰、兰草、兰华、香祖、王者香、空谷仙子、花中君子

科名：兰科

药用：

采集加工　以根或全草入药。四季可采，洗净鲜用或晒干备用。花亦入药。通常用建兰。

性味　辛，平。

功效　滋阴清肺，化痰止咳。

5. Orchid

Scientific name：*Cymbidium*

Alias：Spring Orchid；Cymbidium；Orchis

Family name：Orchidaceae

Collecting and processing for medicine：The whole piece of plant or its roots can be used as medicine. They can be picked in all seasons，washed for fresh use or dried for later use. Flowers are also used as medicine. The species *Cymbidium ensifolium*（L.）Sw. is usually adopted.

Property and flavor：Pungent in flavor，mild in nature.

Efficacy：Enrich *Yin* system and clear away lung-heat，eliminate phlegm and stop coughs.

6. 梅花

学名：*Prunus mume*

别名：春梅、红梅、干枝梅、绿萼梅

科名：蔷薇科

药用：

采集加工　采未成熟的梅子，熏制而成"乌梅"，经盐渍而成"白梅"。花亦入药，以花梅的绿萼梅为好。冬末至次年早春，采摘初放的花朵，摊开晒干即可。梅叶、梅梗和梅根均入药，随采随用或晒干备用，食成熟梅子后收集梅核。

性味　乌梅：酸、涩，温。白梅：酸、涩、咸，平。梅花：微酸，涩，平。梅叶：酸，平。梅梗、梅根：酸、涩。梅核仁：酸，平：有小毒。

功效　乌梅、白梅：敛肺涩肠，除烦，生津止渴，杀蛔、止血。花、根：活血解毒，平肝和胃。

6. Plum

Scientific name：*Prunus mume*

Alias：Spring Plum；Red Plum；Green Calyx Plum

Family name：Rosaceae

Collecting and processing for medicine：Immature plums are picked and then smoked as "dark plums", and later salted to turn into "white plums". Plum flowers are also used as medicine, and the species green calyx plum is regarded as the best. From the end of winter to early spring of the following year, the first blooming flowers will be picked, spread out and dried. Plum leaves, stems, and roots are all used as medicine whenever they are picked or dried for later use. Mature plums are edible, and their kernels can be collected after eating.

Property and flavor：The dark plum is sour and astringent in flavor, mild in nature. The white plum is sour, astringent and salty in flavor, mild in nature. The plum flower is slightly sour, astringent in flavor, and mild in nature. The plum leaf is sour in flavor and mild in nature. The plum stem and the root are sour and astringent in flavor. The plum kernel is sour in flavor, mild and slightly poisonous in nature.

Efficacy：Both dark plum and white plum can astringe the lungs and intestines, relieve restlessness, help produce saliva and quench thirst, kill roundworms and stanch bleeding. The flower and the root can promote blood circulation and remove toxicity, calm the liver and the stomach.

7. 牡丹

学名：*Paeonia suffruticosa*

别名：花王、木芍药、百两金、富贵花、鹿韭、洛阳花

科名：芍药科

药用：

采集加工　栽培3~5年的牡丹，于秋季叶枯萎时挖根，洗净泥土，除须根及茎苗，除去木心，晒干。花初放时采收、晒干。

性味　辛、苦、凉。花：微苦、淡、平。

功效　清热凉血，活血化瘀(制炭用于止血)。

7. Tree Peony

Scientific name：*Paeonia suffruticosa*

Alias：King of Flowers；Ardisia Crispa；Treasure Flower；Luoyang Flower

Family name：Paeoniaceae

Collecting and processing for medicine：When the leaves of peony wilt in autumn, the roots cultivated for 3 to 5 years are dug out, then wash away soil, remove their roots hairs, stems and those that have become lignification, then dry them in the sun. The flowers are picked and dried when they first blossom.

Property and flavor：Pungent and bitter in flavor, cool in nature. The flower is slightly bitter and light in flavor, mild in nature.

Efficacy：Remove pathogenic heat from blood, promote blood circulation to remove stasis (it can be used to make carbon to stanch bleeding).

8. 水仙

学名：*Narcissus tazetta* subsp. *chinensis*

别名：水仙花、天蒜、雅蒜、玉玲珑、金盏银台、金盏玉台、银盏玉台

科名：石蒜科

药用：

采集加工　以鳞茎入药，春、秋季采集，洗去泥沙，用开水烫后、切片晒干或鲜用。

性味　苦、辛，寒。有小毒。

功效　清热解毒，散结消肿，活血通经。

8. Narcissus

Scientific name：*Narcissus tazetta* subsp. *chinensis*

Alias：Daffodil；Narcissus

Family name：Amaryllidaceae

Collecting and processing for medicine：Its bulbs, collected in spring and autumn, are used as medicine. Wash away the soil, scald the bulbs with boiling water, then slice and use them when they are fresh or dry them for later use.

Property and flavor：Bitter, pungent in flavor and cool in nature. Slightly poisonous.

Efficacy：Clear internal heat and remove toxicity, eliminate coagulation and disperse a swelling, promote blood circulation and make menstrual flow smooth.

9. 山茶

学名：*Camellia*

别名：茶花、耐冬、曼陀罗

科名：山茶科

药用:

采集加工　花于含苞待放或初放时采收，晒干或烘干备用，也可鲜用。习惯上以开红花者为好。叶、根全年可采。

性味　苦、微辛，寒。

功效　收敛凉血，止血，调胃，理气。

9. Camellia

Scientific name: *Camellia*

Alias: Japonica; Stramonium

Family name: Theaceae

Collecting and processing for medicine: Flowers are picked when they are in bud or when they first bloom, then dry them for later use or use them when they are fresh. Those with red flowers are better. Leaves and roots can be picked all the year round.

Property and flavor: Bitter, slightly pungent in flavor and cool in nature.

Efficacy: Astringe and remove pathogenic heat from blood, stanch bleeding, nourish the stomach and regulate the flow of *qi*.

10. 月季

学名: *Rosa chinensis*

别名: 胜春、月月红、长春花、斗雪红

科名: 蔷薇科

药用:

采集加工　多以花入药，根、枝叶等亦入药。可于晴天采摘含苞待放的花蕾，及时摊开晒干或用微火烘干。春秋挖根，洗净晒干，叶多鲜用。

性味　甘，温。

功效　活血调经，消肿解毒。

10. Chinese rose

Scientific name: *Rosa chinensis*

Alias: Catharanthus Roseus; Rosa Chinensis; Rose Hybrid

Family name: Rosaceae

Collecting and processing for medicine: Flowers are mostly used as medicine, and the roots, branches and leaves, and soon are also used as medicine. Flower buds can be picked on sunny days; spread them out instantly and dry them in the sun or with a slight fire. Roots are dug in spring and autumn, washed and dried, while the leaves are often used fresh.

Property and flavor: Sweet in flavor and warm in nature.

Efficacy: Promote blood circulation for regulate menstruation, disperse a swelling and remove toxicity.

附录3　常见神经系统病症与花疗用花
Appendix 3　Commonly Used Flowers for Nervous System Symptoms

1. 神经衰弱(百合、含羞草)
1. Neurasthenia (Lily, Mimosa)

(1)百合

学名：*Lilium brownii* var. *viridulum*

别名：山百合、药百合、家百合、喇叭筒

科名：百合科

药用：

采集加工　鳞茎繁殖2年后秋季采收，洗净，剥取鳞片，用开水烫或蒸5~10分钟，至百合边缘柔软或背面有极小的裂纹时，迅速取出，用清水洗净去黏液，晒干备用。

性味　甘，平。花：甘、微苦，微寒，平。

功效　润肺止咳，清心安神，通利大小便。

(1)Lily

Scientific name：*Lilium brownii* var. *viridulum*

Alias：Lilium Auratum；Lilium Speciosum

Family name：Liliaceae

Collecting and processing for medicine：The bulbs of lily are collected in autumn after two years of breeding, washed and peeled off the scales, scalded with hot water or steamed for 5-10 minutes until the edges of the lily are soft or there are very small cracks on the back, then quickly take them out and wash off the mucus, dry them for later use.

Property and flavor：The flowers are sweet and slightly bitter in flavor, slightly cold and mild in nature.

Efficacy：Moisten lung for stopping coughs, clear mind of worries and make soothing, and relax the bowels.

(2)含羞草

学名：*Mimosa pudica*

别名：知羞草、怕羞草、怕丑草、感应草、抓痒花

科名：豆科

药用：

采集加工　夏秋采集全草，洗净，切段，晒干备用。

性味　甘、涩，微寒。有小毒。

功效　安神镇静，散瘀止痛，止血收敛。

（2）**Mimosa**

Scientific name：*Mimosa pudica*

Alias：Sensitive plant；Shameplant；Touch-Me-Not；Ticklish Plant

Family name：Fabaceae

Collecting and processing for medicine：The whole plant is picked in summer and autumn, then washed and cut into pieces, dried for later use.

Property and flavor：Sweet and astringent in flavor, lightly cold in nature. Slightly poisonous.

Efficacy：Make soothing, remove stasis to relieve pains, stanch bleeding to astringe.

2. 健忘多梦(合欢花)
2. Forgetfulness and Dreaminess(Silk Tree)

合欢花

学名：*Albizzia julibrissin*

别名：绒花树、夜合树、芙蓉花树、马缨花

科名：豆科

药用：

采集加工　夏秋锯细干，剥取皮切片，晒干。花在初开时，择晴天摘下，迅速晒干或放阴处晾干，需注意经常翻动；花蕾(商品名称为"合欢米")在 5 月花未开时采收，晒干。

性味　甘、平

功效　合欢皮：安神解郁、和血止痛。合欢花、合欢米：理气、解郁、开胃、养心、活络、养心安神。

Silk Tree

Scientific name：*Albizzia julibrissin*

Alias：Mimosa Tree

Family name：Fabaceae

Collecting and processing for medicine：The thin branches of the plant are sawn in summer and autumn, with their barks peeled, then sliced into pieces and dried in the sun. When the flowers firstly bloom, pick them on a sunny day, and dry them in the sun quickly or in the shade, and turn them over frequently；Flower buds (commercially known as "Albizia Durazz") are picked in May before the flowers bloom and dried in the sun.

Property and flavor：Sweet and neutral

Efficacy：The cortex albiziae can clear mind of worries and make soothing, and regulate blood circulation to alleviate pains；the flowers and the product *Albizia Durazz* can regulate the flow of *qi*, relieve emotional distress, stimulate the appetite, improve collateral circulation, nourish the heart and smooth the nerves.

3. 偏头痛(向日葵)
3. Migraine(Sunflower)

　　向日葵

　　学名：*Helianthus anns*

　　别名：葵花、向阳花、朝阳花、转莲(苏北)、照葵(鲁南)

　　科名：菊科

　　药用：

　　采集加工　可在籽成熟后连根拔起，分别采收籽、籽壳、花盘、茎髓、根；花瓣在初开时采收；叶多鲜用，也可晒干备用。

　　性味　籽：淡，平。花盘：甘，温。花、叶：苦。

　　功效　籽：润肺平肝，祛风除湿，驱虫利尿。花盘：降压止痛，养肝补肾，茎髓、根：清热利尿、止咳平喘。花瓣、叶：健胃。

Sunflower

Scientific name：*Helianthus anns*

Alias：Tournesol; Zhuanlian (northern Jiangsu in China); Zhaokui (southern Shandong in China)

Family name：Asteraceae

Collecting and processing for medicine：The plant is uprooted after its seeds are ripe, then its seeds, shells, flower discs, stem pith and roots are separately picked and collected; the petals are collected when the flowers first blossom; the leaves are more used fresh and can also be dried for later use.

Property and flavor：Seeds are light in taste, and mild in nature. Flower discs are sweet, warm in nature. Flowers and the leaves are bitter.

Efficacy：Seeds can moisten the lungs and calm the liver, resist external wind-evil and expel internal dampness, expel parasite and help urinate. Flower discs can lower blood pressure and relieve pains, nourish the liver and tonify the kidney. Stem pith and roots can expel internal heat and help urinate, relieve cough and asthma. Petals and leaves can nourish stomach.

4. 平日忧愁(萱草)
4. Daily annoyances and worries (Daylily)

　　萱草

　　学名：*Hemerocallis fulva*

　　别名：黄花菜、金针菜、宜男花、忘忧

　　科名：阿福花科

　　药用：

采集加工　嫩苗春天采收，多鲜用；花在未开放或初放时采收，蒸熟后晒干；秋冬挖根，洗净，开水烫后晒干。

性味　甘，凉。根有毒。

功效　花为营养蔬菜，嫩苗亦可食。二者清热解毒，止血，止渴生津，利尿，解酒毒。根为强壮滋补药，清热解毒，利尿消肿。

Daylily

Scientific name：*Hemerocallis fulva*

Alias：Dried Lily；Forgetting Sorrow

Family name：Asphodelaceae

Collecting and processing for medicine：The plant's tender shoots are picked in spring, and used fresh more often；the flowers are picked and collected when they are in bud or first blossom, then steamed and dried in the sun；the roots are dug in autumn and winter, then washed and scalded with boiling water, and finally dried for use.

Property and flavor：Sweet in flavor, cool in nature. The roots are poisonous.

Efficacy：The flowers are nutrient vegetables, and the tender shoots are also edible. Both can clear internal heat and remove toxicity, stanch bleeding, quench thirst and promote saliva secretion, help urinate, dispel the effects of alcohol. The root is a tonic for strengthening body. It can clear internal heat and remove toxicity, help urinate and disperse a swelling.

5. 忧郁、烦闷（金橘）

5. Moping and Depression（Kumquat）

金橘

学名：*Citrus japonica*

别名：金谈、罗浮、金枣、牛奶金柑、羊奶橘

科名：芸香科

药用：

采集加工　果实成熟时采收，取果皮和络、核晒干备用。橘皮的外层红色薄皮，叫"橘红"；内层的白皮叫"橘白"；未成熟的果实或青色果皮，称为"青皮"，可取自然落下的或淘汰的幼果，洗净晒干。叶可随采随用。

性味　橘皮：苦、辛，温。橘络、橘红、橘白：苦，平。橘核：苦，湿。橘叶：苦，平。青皮：苦、辛，温。

功效　橘皮：温气、健胃、燥湿、化痰。橘络：通络、化痰。橘核：理气、散结、止疼。橘红、橘白：止咳、化痰、理气。青皮：破气、散结、疏肝、止痛、消食化滞。橘叶：疏肝利气。

Kumquat

Scientific name：*Citrus japonica*

Alias：Calamondin Orange；Cumquat

Family name：Rutaceae

Collecting and processing for medicine：Kumquat fruits are picked when they are ripe, with their peels, tangerine pith and cores dried for later use. The outer red thin peel is called "orange redness"; the white one on the inner layer is called "orange whiteness"; the immature fruit or green peel is called "green orange peel". Naturally fallen or eliminated immature fruits can be washed and dried for use. Leaves can be picked and used at any time.

Property and flavor：The peels are bitter and pungent, warm in nature. The tangerine pith, the orange redness and the whiteness are all bitter flavor, mild in nature. The cores are bitter, humid. The leaves are bitter flavor, mild in nature. The green orange peels are bitter and pungent, warm in nature.

Efficacy：The peels can warm up qi, nourish stomach, eliminate internal dampness and resolve phlegm. The tangerine pith can dredge collaterals, resolve phlegm. The cores can regulate the flow of qi, remove stasis and relieve pains. The orange redness and the whiteness can relieve cough, resolve phlegm, and regulate the flow of qi. The green orange peel can relieve unsmooth flow of qi, remove stasis, sooth the liver, relieve pains and help digest. The leaves can sooth the liver and regulate the flow of qi.

6. 心悸、烦燥(景天)
6. Palpitation and Irritability (Sedum)

景天
学名：*Hylotelephium erythrostictum*
别名：八宝、蝎子草、土三七、护花草、慎火
科名：景天科
药用：
采集加工　以全草及根入药。全年可采，鲜用或晒干备用。
性味　苦、酸，寒。
功效　祛风清热，活血化瘀，止血止痛。

Sedum

Scientific name：*Hylotelephium erythrostictum*

Alias：Hylotelephium Erythrostictum; Nilgiri Nettle; Sedum Uizoon

Family name：Crassulaceae

Collecting and processing for medicine：The whole plant and its roots can be used as medicine. It can be picked all the year, used fresh or dried for later use.

Property and flavor：Bitter and sour, cool in nature.

Efficacy：Resist external wind-evil and clear internal heat, promote blood circulation to dissipate stasis, stop bleeding and relieve pains.

7. 失眠(荷包牡丹)
7. Insomnia(Bleeding-heart)

荷包牡丹

学名：*Lamprocapnos spectabilis*

别名：鱼儿牡丹、兔儿牡丹、玲儿草

科名：罂粟科

药用：

采集加工　秋季或春季采根，洗净晒干备用。

性味　辛，温。

功效　散血、镇痛、除风、和血、消疮毒。

Bleeding-heart

Scientific name：*Lamprocapnos Spectabilis*

Alias：Common Bleeding Heart；Colicweed

Family name：Papaveraceae

Collecting and processing for medicine：The roots are collected in autumn or spring, then washed and dried for later use.

Property and flavor：Pungent, warm in nature.

Efficacy：Remove blood stasis, relieve pains, resist external wind-evils, regulate blood circulation and eliminate sore poison.